THE

LAWS OF GRAVITATION

MEMOIRS BY NEWTON, BOUGUER AND CAVENDISH

TOGETHER WITH ABSTRACTS OF OTHER IMPORTANT MEMOIRS

TRANSLATED AND EDITED BY

A. STANLEY MACKENZIE, Ph.D.

PROFESSOR OF PHYSICS IN BRYAN MAWR COLLEGE

**University Press of the Pacific
Honolulu, Hawaii**

The Laws of Gravitation:
Memoirs by Newton, Bouguer and Cavendish

Translated and Edited by
A. Stanley Mackenzie

ISBN: 1-4102-0254-2

Reprinted from the 1900 edition

University Press of the Pacific
Honolulu, Hawaii
http://www.universitypressofthepacific.com

GENERAL CONTENTS

	PAGE
Preface...	v
History of the subject before the appearance of Newton's *Principia*.	1
Extracts from Newton's *Principia* and *System of the World*........	9
Biographical sketch of Newton.......	19
Bouguer's *The Figure of the Earth*	23
Biographical sketch of Bouguer................................ ..	44
The Bertier controversy....................	47
Account of Maskelyne's experiments on Schehallien...............	53
Cavendish's *Experiments to determine the mean density of the Earth*...	59
Biographical sketch of Cavendish....................	107
Historical account of the experiments made since the time of Cavendish..	111
Table of results of experiments...................................	143
Bibliography..	145
Index...	157

PREFACE

In preparing this volume, the ninth in the Scientific Memoirs series, the editor has had in mind the fact that the most important of the memoirs here dealt with, that of Cavendish, is frequently given for detailed study to young physicists in order to train them in the art of reading for themselves periodical scientific literature.   Certainly no better piece of work could be used for the purpose, whether one considers the intrinsic importance of the subject-matter, the keenness of argument and the logical presentation in detail, or the use and design of apparatus and the treatment of sources of error. The main objections to Cavendish's work are those he himself pointed out, and it is important to notice that, notwithstanding all the advance in the refinement and manipulation of apparatus which has been made during the century that has elapsed since the date of Cavendish's experiment, his value for the mean specific gravity of the earth, 5.448, must still be considered one of the most reliable, being not far from the latest results of Poynting, König and Richarz and Krigar-Menzel, Boys and Braun.

Believing that we in America devote insufficient time, if any, to a study of Newton's great work, the editor has thought it well to incorporate with the memoirs on the experimental investigation of gravitational attraction the statements of Newton himself concerning that subject.

The laws of gravitation are embodied in the formula,

$$f = G\frac{mm'}{r^2},$$

which says that the attraction between two particles of matter is directly proportional to the product of their masses, inversely proportional to the square of the distance between them, and independent of the kind of matter and of the intervening medium. G is then a constant in nature, the Gravitation Constant. It is more common perhaps to speak of the law, than of the laws, of gravitation ; this has no doubt arisen from the fact that they can be stated in a single mathematical formula. The best evidence of the truth of these laws is indirect, for, assuming them valid, astronomical measurements show that they account for all the motions of the heavenly bodies. Such measurements do not, however, enable us to find the numerical value of G ; for that purpose we must determine the attraction between two masses of known amount at a known distance apart. It is with experiments of this character that the present volume has to deal. As the masses used in such experiments vary from a metal sphere of a few tenths of an inch in diameter to a huge mountain mass, or to a shell of the earth's crust 1250 ft. in thickness, and as the attraction has been observed with such different instruments as the plumb-line, the pendulum, the torsion balance, the pendulum balance and the beam balance, and yet the resulting value of G is always about the same, we can regard these experiments as constituting a further proof of Newton's laws, and the editor has accordingly felt justified in using the title given. Assuming the earth to be a sphere, the value of G is connected with the value of the mean specific gravity of the earth, Δ, by the equation

$$G\Delta = \frac{3g}{4\pi R},$$

where g is the acceleration due to gravity, and R the radius of the earth ; and, accordingly, it is quite usual to state that the aim of the above experiments is to find the mean density of the earth.

The work on the attraction of mountain masses by the French Academicians Bouguer and de la Condamine in Peru is of very great importance, and is not known as it deserves to be ; almost all of their account of the work is therefore here presented. It will be seen that they were the pioneers in two of the methods which have been used for the measurement of gravitational attraction ; and although, on account of imperfect instruments and unfavourable local conditions, their numerical results are untrustworthy, they give the theory and method of the experiments with great originality and clearness. Such notes have been added to the memoirs as seemed

PREFACE

necessary to prevent the reader from wasting time over obscure and inaccurate passages, and to suggest material for collateral reading.

An effort has been made to present along with the memoirs a brief historical account of the various modes of experiment used for finding the mean specific gravity of the earth, and a table of results is added. As the literature on the subject before the present century is not always easily obtainable, the treatment of the matter for that period is given in comparatively greater detail. Believing that a bibliography containing every important reference to the subject is an essential feature of a work of this kind, the editor has endeavoured to make himself familiar with the whole of the very extensive literature relating to it, and accordingly is fairly confident that no important memoir has escaped his observation. From the mass of material thus collected the bibliography given at the end of the volume has been compiled. In order to keep within the limits of space assigned, some references had to be omitted, but they relate mainly to recent work, and it is believed that they contain nothing of importance.

No effort has been made to deal with the mathematical side of the subject; accordingly the memoirs of Laplace, Legendre, Ivory, etc., which deal with the finding of the mean specific gravity of the earth by means of analytical methods are not referred to; but it is hoped that all the more important experimental investigations have been touched upon.

A. STANLEY MACKENZIE.

BRYN MAWR, *October*, 1899.

vii

HISTORY OF THE SUBJECT
BEFORE THE
APPEARANCE OF NEWTON'S "PRINCIPIA"

Dr. Gilbert's contributions to the speculations on gravitation are among the most important of the early writings on that subject, although to Kepler also must credit be given for a deep insight into its nature; the latter announces in his introduction to the *Astronomia Nova*, published in 1609, his belief in the perfect reciprocity of the action of gravitation, and in its application to the whole material universe. Gilbert was led by his researches on magnetism to the conclusion that the force of gravity was due to the magnetic properties of the earth; and in 1600 announced [1*, I, 21] his opinion that bodies when removed to a great distance from the earth would gradually lose their motion downwards. The earliest proposals we find for investigating whether such changes occur in the force of gravity are in the works of Francis Bacon [2, *Nov. Org.* II, 36, and *Hist. Nat.* I, 33]. He maintained that this force decreased both inwards and outwards from the surface of the earth, and suggested experiments to test his views. He would take two clocks, one actuated by weights and the other by the compression of an iron spring, and regulate them so that they would run at the same rate. The clock actuated by weights was then to be placed at the top of some high steeple, and at the bottom of a mine, and its rate at each place compared with that of the other, which remained at the surface. There is no record of any trial of the experiment at that time.

After the founding of the Royal Society of London a stimulus was given to experimenting upon this as upon many other

---

* The numbers in brackets refer to the Bibliography.

subjects. A paper was read before that society by Dr. Power [10, vol. 1, p. 133] on December 3, 1662, upon "Subterraneous Experiments." A pound weight and 68 yards of thread were put into one pan of a scale and counterpoised. The weight was then lowered into a pit and attached by means of the thread to the scale-pan held directly over the mouth of the pit ; it was found to lose in weight by at least an ounce.* Three weeks later Hooke [10, vol. 1, p. 163] made a report to the society on some experiments he had performed at Westminster Abbey. The report is worth reprinting, as giving some idea of the method employed in such experiments, and of the state of knowledge upon the subject at the time when Newton first took it up. For it will be remembered that it was in 1665 that Newton was led by his speculations on gravity to imagine that since this action did not sensibly diminish with small changes in height, it might perhaps extend to the moon, and be the cause of that body's being retained in her orbit. Pursuing his train of thought, he extended this explanation to the sun and planets: and, taking into consideration Kepler's laws, it was therefore necessary that the force must fall off in the inverse ratio of the square of the distance. When applying this law of decrease from the earth to the moon, Newton used in deducing the length of the radius of the earth the rough estimate, then current, of 60 miles to a degree of latitude, instead of nearly 69½ ; and as a consequence the calculated motion of the moon did not agree with the observed motion. He thereupon laid aside for the time being any further thought upon the matter. His attention was again called to the subject by a letter from Hooke in 1679, and, Picard having in the meantime measured the earth, Newton was able to apply the correct data to the problem and to arrive at a beautiful agreement of the calculated with the observed behaviour of the moon. From that time date the wonderful researches which were the foundation of the *Principia.* The following is Hooke's report :

"In prosecution of my Lord Verulam's experiment concerning the decrease of gravity, the farther a body is removed below the surface of the earth, I made trial, whether any such difference in the weight of bodies could be found by their

* According to Le Sage [34], Descartes had suggested a similar undertaking twenty-five years earlier in a letter to Mersenne.

nearer or farther removal from that surface upwards. To this end I took a pair of exact scales and weights, and went to a convenient place upon Westminster Abbey, where was a perpendicular height above the leads of a subjacent building, which by measure I found threescore and eleven foot. Here counterpoising a piece of iron (which weighed about 15 ounces troy) and packthread enough to reach from the top to the bottom, I found the counterpoise to be of troy-weight seventeen ounces and thirty grains. Then letting down the iron by the thread, till it almost touched the subjacent leads, I tried what alteration there had happened to its weight, and found, that the iron preponderated the former counterpoise somewhat more than ten grains. Then drawing up the iron and thread with all the diligence possibly I could, that it might neither get nor lose any thing by touching the perpendicular wall, I found by putting the iron and packthread again into its scale, that it kept its last equilibrium; and therefore concluded, that it had not received any sensible difference of weight from its nearness to or distance from the earth. I repeated the trial in the same place, but found, that it had not altered its equilibrium (as in the first trial) neither at the bottom, nor after I had drawn it up again ; which made me guess, that the first preponderating of the scale was from the moisture of the air, or the like, that had stuck to the string, and so made it heavier. In pursuance of this experiment, I removed to another place of the Abbey, that was just the same distance from the ground, that the former was from the leads; and upon repeating the trial there with the former diligence, I found not any sensible alteration of the equilibrium, either before or after I had drawn it up; which farther confirmed me, that the first alteration proceeded from some other accident, and not from the differing gravity of the same body.

"I think therefore it were very desirable, from the determination of Dr. Power's trials, wherein he found such difference of weight, that it were examined by such as have opportunity, first, what difference there is in the density and pressure of the air, and what of that condensation of gravity may be ascribed to the differing degrees of heat and cold at the top and bottom, which may be easily tried with a common weatherglass and a sealed-up thermometer ; for the thermometer will shew what of the change is to be ascribed to heat and cold, and the weather-glass will shew the differing condensation. Next, for the knowing, whether this alteration of gravity proceed from the density and gravity of the ambient air, it would be requisite to make use of some very light body, extended into large dimensions, such as a large globe of glass carefully stopt, that no air may get in or out; for if the alteration proceeded from the magnetical attraction of the parts of the earth, the ball will lose but a sixteenth part of its weight (supposing a lump of glass held the same proportion, that Dr. Power found in brass); but if it proceed from the density of the air, it may lose half, or perhaps more. Further, it were very desireable, that the current of the air in that place were observed, as Sir Robert Moray intimated the last day. Fourthly, I think it were worth trial to counterpoise a light and heavy body one against another above, and to carry down the scales and them to the bottom, and observe what happens. Fifthly, it were desireable, that trials were made, by the letting down of other both heavier and lighter bodies, as lead, quicksilver, gold, stones, wood, liquors, animal substances, and the like. Sixthly, it were to be wished, that trial were made how that gravitation does decrease with the descent of the body—that is, by making trial, how much the body grows lighter at every ten or twenty foot distance. These trials, if accurately made, would afford a great help to guess at the cause of this strange phaenomenon."

Dr. Power's experiment was repeated by Dr. Cotton, and an account of his trials was given to the Society on June 1, 1664 [10, vol. 1, p. 433]. The weight was $\frac{1}{4}$ lb., and the length of the string 36 yards. A loss in weight of $\frac{1}{4}$ oz. was found.

On September 1, 1664 [6, vol. 5, p. 307], we find a reference to some experiments made at St. Paul's Cathedral by a committee of the Royal Society consisting of Sir R. Moray, Dr. Wilkins, Dr. Goddard, Mr. Palmer, Mr. Hill and Mr. Hooke. The results of these experiments were given to the Society on September 14, 1664 [10, vol. 1, p. 466] ; the weight was 15 lbs. troy, the string about 200 ft. long, and the loss of weight 1 drachm. In a letter to Mr. Boyle [6, vol. 5, p. 536], dated September 15th, Mr. Hooke gives more details, and remarks that the balance was sensitive enough to be turned by a few grains. He suggests the variation of the density of the air as the cause of the loss in weight. Boyle [10, vol. 1, p. 470] proposed that Hooke's suggestion be tested by making the suspended weight of a large glass ball loaded with mercury.

At a meeting of the Royal Society on March 14, 1665, Hooke reported [10, vol. 2, p. 66, and 6, vol. 5, p. 544] that he had tried Dr. Power's experiment at some wells near Epsom and had found no loss in weight. Similar experiments were made by Hooke at Banstead Downs, in Surrey, and reported on March 21, 1666 [10, vol. 2, p. 69, and 6, vol. 5, pp. 355 and 546]. The string was 330 ft. long, and the balance sensitive to a grain, yet a pound shewed no change in weight when suspended at the bottom of the well. He concludes that the power of gravity cannot be magnetical, as Gilbert had supposed. He says: " But in truth upon the consideration of the nature of the theory, we may find, that supposing it true, that all the constituent parts of the earth had a magnetical power, the decrease of gravity would be almost a hundred times less than a grain to a pound, at as great a depth as fifty fathom ; for if we consider the proportion of the parts of the earth placed upon one side beneath the stone, with the parts on the other side above it, we may find the disproportion greater. Unless we suppose the magnetism of the parts to act but at a very little distance, which I think the experiments made in the Abbey and St. Paul's will not allow of. If therefore there be any such inequality of gravity, we must have some ways of trial much more accurate than this of scales, of which I shall propound two sorts," etc. It is interesting to notice that the considerations upon which he makes his computations are practically those used by Airy in his Harton Colliery experiment.

On December 7, 1681 [10, vol. 4, p. 110], Hooke produced before the society two pendulum-clocks adjusted to run at the same rate. He proposed to put one at the top and the other at the bottom of the monument on Fish Street Hill, and observe whether they would keep ·together. No notice of his having tried the experiment has been found. This is the method proposed by Bacon and used by Boùguer and many others.

In 1682, Hooke read before the Royal Society " A Discourse of the Nature of Comets " [4, pp. 149–191], in which he gives his ideas on the subject of gravity (particularly on pages 170–183). He considers gravity to be a universal principle, inherent in all matter, propagated by the same medium as that by means of which light is conveyed, with unimaginable celerity, to indefinitely great distances, and with a power varying with the distance. He sums up his conceptions on gravitation in nine propositions, which are of great interest, in that they include many of the conceptions of Newton on this subject, and yet were published four years before the *Principia* appeared.

6

PHILOSOPHIAE NATURALIS PRINCIPIA MATHEMATICA

1st Edition, London, 1687   2d Edition, Cambridge, 1713 (Cotes' Edition)
3d Edition, London, 1726 (Pemberton's Edition)

AND

DE MUNDI SYSTEMATE

London, 1727

BY

SIR ISAAC NEWTON

(Extracts taken from Davis's Edition of Motte's translation
3 volumes, London, 1803)

CONTENTS

		PAGE
On the attraction of spheres		9
Law of the distance		9
Law of the masses		13
Variation of gravity on the earth's surface		14
All attraction is mutual		15
Methods of showing the attraction between terrestrial bodies		17
Proof of its existence		17
Similar discussion for the case of celestial bodies		18
Final statements concerning the laws of gravitation		19

8

THE MATHEMATICAL PRINCIPLES OF NATURAL PHILOSOPHY

AND

SYSTEM OF THE WORLD

BY

SIR ISAAC NEWTON

BOOK I. PROPOSITION LXXIV. THEOREM XXXIV.

The same things supposed (if to the several points of a given sphere there tend equal centripetal forces decreasing in a duplicate ratio of the distances from the points), *I say, that a corpuscle situate without the sphere is attracted with a force reciprocally proportional to the square of its distance from the centre.*

BOOK I. PROPOSITION LXXV. THEOREM XXXV.

If to the several points of a given sphere there tend equal centripetal forces decreasing in a duplicate ratio of the distances from the points; I say, that another similar sphere will be attracted by it with a force reciprocally proportional to the square of the distance of the centres.

For the attraction of every particle is reciprocally as the square of its distance from the centre of the attracting sphere (by prop. 74), and is therefore the same as if that whole attracting force issued from one single corpuscle placed in the centre of this sphere. But this attraction is as great as on the other hand the attraction of the same corpuscle would be, if that were itself attracted by the several particles of the attracted sphere with the same force with which they are attracted by it. But that attraction of the corpuscle would be (by prop. 74) reciprocally proportional to the square of its distance from the centre of the sphere; therefore the attraction of the sphere, equal thereto, is also in the same ratio. Q. E. D.

Cor. 1. The attractions of spheres towards other homogeneous spheres are as the attracting spheres applied to the squares of the distances of their centres from the centres of those which they attract.

Cor. 2. The case is the same when the attracted sphere does also attract. For the several points of the one attract the several points of the other with the same force with which they themselves are attracted by the others again; and therefore since in all attractions (by law 3) the attracted and attracting point are both equally acted on, the force will be doubled by their mutual attractions, the proportions remaining.

[*Proposition LXXVI. proves the same thing for spheres made up of homogeneous concentric layers.*]

BOOK III. PROPOSITION V. THEOREM V. SCHOLIUM.

The force which retains the celestial bodies in their orbits has been hitherto called centripetal force; but it being now made plain that it can be no other than a gravitating force, we shall hereafter call it gravity. For the cause of that centripetal force which retains the moon in its orbit will extend itself to all the planets.

BOOK III. PROPOSITION VI. THEOREM VI.

That all bodies gravitate towards every planet; and that the weights of bodies towards any the same planet, at equal distances from the centre of the planet, are proportional to the quantities of matter which they severally contain.

It has been, now of a long time, observed by others, that all sorts of heavy bodies (allowance being made for the inequality of retardation which they suffer from a small power of resistance in the air) descend to the earth *from equal heights* in equal times; and that equality of times we may distinguish to a great accuracy, by the help of pendulums. I tried the thing in gold, silver, lead, glass, sand, common salt, wood, water, and wheat. I provided two wooden boxes, round and equal;

I filled the one with wood, and suspended an equal weight of gold (as exactly as I could) in the centre of oscillation of the other. The boxes hanging by equal threads of 11 feet made a couple of pendulums perfectly equal in weight and figure, and equally receiving the resistance of the air. And, placing the one by the other, I observed them to play together forwards and backwards, for a long time, with equal vibrations. . . . and the like happened in the other bodies. By these experiments, in bodies of the same weight, I could manifestly have discovered a difference of matter less than the thousandth part of the whole, had any such been. But, without all doubt, the nature of gravity towards the planets is the same as towards the earth. . . . Moreover, since the satellites of Jupiter perform their revolutions in times which observe the sesquiplicate proportion of their distances from Jupiter's centre, their accelerative gravities towards Jupiter will be reciprocally as the squares of their distances from Jupiter's centre—that is, equal at equal distances. And, therefore, these satellites, if supposed to fall *towards Jupiter* from equal heights, would describe equal spaces in equal times, in like manner as heavy bodies do on our earth. . . . If, at equal distances from the sun, any satellite, in proportion to the quantity of its matter, did gravitate towards the sun with a force greater than Jupiter in proportion to his, according to any given proportion, suppose of d to e; then the distance between the centres of the sun and of the satellite's orbit would be always greater than the distance between the centres of the sun and of Jupiter nearly in the subduplicate of that proportion; as by some computations I have found. And if the satellite did gravitate towards the sun with a force, lesser in the proportion of e to d, the distance of the centre of the satellite's orbit from the sun would be less than the distance of the centre of Jupiter from the sun in the subduplicate of the same proportion. Therefore if, at equal distances from the sun, the accelerative gravity of any satellite towards the sun were greater or less than the accelerative gravity of Jupiter towards the sun but by one $\frac{1}{1000}$ part of the whole gravity, the distance of the centre of the satellite's orbit from the sun would be greater or less than the distance of Jupiter from the sun by one $\frac{1}{2000}$ part of the whole distance— that is, by a fifth part of the distance of the utmost satellite from the centre of Jupiter; an eccentricity of the orbit which would be very sensible. But the orbits of the satellites are concentric to Jupiter, and therefore the accelerative gravities of Jupiter, and of all its satellites towards the sun, are equal among themselves. . . .

But further; the weights of all the parts of every planet towards any other planet are one to another as the matter in the several parts; for if some parts did gravitate more, others less, than for the quantity of their matter, then the whole planet, according to the sort of parts with which it most abounds, would gravitate more or less than in proportion to the quantity of matter in the whole. Nor is it of any moment whether these parts are external or internal; for if, for example, we should imagine the terrestrial bodies with us to be raised up to the orb of the moon, to be there compared with its body; if the weights of such bodies were to the weights of the external parts of the moon as the quantities of matter in the one and in the other respectively; but to the weights of the internal parts in a greater or less proportion, then likewise the weights of those bodies would be to the weight of the whole moon in a greater or less proportion; against what we have shewed above.

Cor. 1. Hence the weights of bodies do not depend upon their forms and textures; for if the weights could be altered with the forms, they would be greater or less, according to the variety of forms, in equal matter; altogether against experience.

Cor. 2. Universally, all bodies about the earth gravitate towards the earth; and the weights of all, at equal distances from the earth's centre, are as the quantities of matter which they severally contain. This is the quality of all bodies within the reach of our experiments; and therefore (by rule 3) to be affirmed of all bodies whatsoever. . . .

Cor. 5. The power of gravity is of a different nature from the power of magnetism; for the magnetic attraction is not as the matter attracted. Some bodies are attracted more by the magnet; others less; most bodies not at all. The power of magnetism in one and the same body may be increased and diminished; and is sometimes far stronger, for the quantity of matter, than the power of gravity; and in receding from the magnet decreases not in the duplicate but almost in the triplicate proportion of the distance, as nearly as I could judge from some rude observations.

THE LAWS OF GRAVITATION

Book III. Proposition VII. Theorem VII.

That there is a power of gravity tending to all bodies, proportional to the several quantities of matter which they contain.

That all the planets mutually gravitate one towards another, we have proved before ; as well as that the force of gravity towards every one of them, considered apart, is reciprocally as the square of the distance of places from the centre of the planet. And thence (by prop. 69, book I, and its corollaries) it follows, that the gravity tending towards all the planets is proportional to the matter which they contain.

Moreover, since all the parts of any planet A gravitate towards any other planet B ; and the gravity of every part is to the gravity of the whole as the matter of the part to the matter of the whole ; and (by law 3) to every action corresponds an equal reaction ; therefore the planet B will, on the other hand, gravitate towards all the parts of the planet A ; and its gravity towards any one part will be to the gravity towards the whole as the matter of the part to the matter of the whole. Q. E. D.

Cor. 1. Therefore the force of gravity towards any whole planet arises from, and is compounded of, the forces of gravity towards all its parts. Magnetic and electric attractions afford us examples of this ; for all attraction towards the whole arises from the attractions towards the several parts. The thing may be easily understood in gravity, if we consider a greater planet as formed of a number of lesser planets meeting together in one globe ; for *hence it would appear that* the force of the whole must arise from the forces of the component parts. If it is objected that, according to this law, all bodies with us must mutually gravitate one towards another, I answer, that since the gravitation towards these bodies is to the gravitation towards the whole earth as these bodies are to the whole earth, the gravitation towards them must be far less than to fall under the observation of our senses.

Cor. 2. The force of gravity towards the several equal particles of any body is reciprocally as the square of the distance of places from the particles ; as appears from cor. 3, prop. 74, book I.

[*Under proposition X occurs the following important passage :*] However the planets have been formed while they were yet in fluid masses, all the heavier matter subsided to the centre.

Since, therefore, the common matter of our earth on the surface thereof is about twice as heavy as water, and a little lower, in mines, is found about three, or four, or even five times more heavy, it is probable that the quantity of the whole matter of the earth may be five or six times greater than if it consisted all of water.*

[*Under propositions XVIII. and XIX., Newton proves that the axes of the planets are less than the diameters drawn perpendicular to the axes. He shows how centrifugal force acts in determining the form of the earth, and discusses the measurements of terrestrial arcs known at that time; he deduces therefrom that gravity will be lessened at the equator by $\frac{1}{230}$ of itself, and that the earth will be higher at the equator than at the poles by 17.1 miles.*]

BOOK III. PROPOSITION XX. PROBLEM IV.

To find and compare together the weights of bodies in the different regions of our earth.

Because the weights of the unequal legs of the canal of water ACQ*qca* are equal ; and the weights of the parts proportional

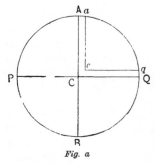

Fig. a

to the whole legs, and alike situated in them, are one to another as the weights of the wholes, and therefore equal betwixt themselves; the weights of equal parts, and alike situated in the legs, will be reciprocally as the legs—that is, reciprocally as 230 to 229. And the case is the same in all homogeneous equal bodies alike situated in the legs of the canal. Their weights are reciprocally as the legs—that is, reciprocally as the distances of the bodies from the centre of the earth. Therefore, if the bodies are situated in the uppermost parts of the canals, or on the surface of the earth, their weights will be one to another reciprocally as their distances from the centre. And, by the same argument, the weights in all other places round the whole surface of the earth are reciprocally as the distances of

* [*This was a wonderfully good guess on Newton's part, since the best of the later determinations give about 5.5 for the mean specific gravity of the earth.*]

the places from the centre; and, therefore, in the hypothesis of the earth's being a spheroid, are given in proportion.

[*Newton then states that " the lengths of pendulums vibrating in equal times are as the forces of gravity"; he enumerates the experiments on the periods of pendulums made at different parts of the earth's surface, and tests his conclusions.*

The following remarks appear on pp. 20–25 of Motte's translation of the " de Mundi Systemate," wherein Newton, after a reference to his pendulum experiments, given on p. 11 of this volume, says:]

Since the action of the centripetal force upon the bodies attracted is, at equal distances, proportional to the quantities of matter in those bodies, reason requires that it should be also proportional to the quantity of matter in the body attracting.

For all action is mutual, and (by the third law of motion) makes the bodies mutually to approach one to the other, and therefore must be the same in both bodies. It is true that we may consider one body as attracting, another as attracted; but this distinction is more mathematical than natural. The attraction is really common of either to other, and therefore of the same kind in both.

And hence it is that the attractive force is found in both. The sun attracts Jupiter and the other planets; Jupiter attracts its satellites; and, for the same reason, the satellites act as well one upon another as upon Jupiter, and all the planets mutually one upon another.

And though the mutual actions of two planets may be distinguished and considered as two, by which each attracts the other, yet, as those actions are intermediate, they do not make but one operation between two terms. Two bodies may be mutually attracted each to the other by the contraction of a cord interposed. There is a double cause of action, to wit, the disposition of both bodies, as well as a double action in so far as the action is considered as upon two bodies; but as betwixt two bodies it is but one single one. It is not one action by which the sun attracts Jupiter, and another by which Jupiter attracts the sun; but it is one action by which the sun and Jupiter mutually endeavour to approach each the other. By the action with which the sun attracts Jupiter, Jupiter and the sun endeavour to come nearer together (by the third law of motion); and by the action with which Jupiter attracts the sun, likewise Jupiter and the sun endeavour to come nearer together. But the sun is not attracted towards Jupiter by a twofold action, nor Jupiter by a twofold action towards the sun ; but it is one single intermediate action, by which both approach nearer together.

Thus iron draws the loadstone as well as the loadstone draws the iron ; for all iron in the neighbourhood of the loadstone draws other iron. But the action betwixt the loadstone and iron is single, and is considered as single by the philosophers. The action of iron upon the loadstone is, indeed, the action of the loadstone betwixt itself and the iron, by which both endeavour to come nearer together ; and so it manifestly appears, for if you remove the loadstone the whole force of the iron almost ceases.

In this sense it is that we are to conceive one single action to be exerted betwixt two planets, arising from the conspiring natures of both ; and this action standing in the same relation to both, if it is proportional to the quantity of matter in the one, it will be also proportional to the quantity of matter in the other.

Perhaps it may be objected that, according to this philosophy (prop. 74, book I), all bodies should mutually attract one another, contrary to the evidence of experiments in terrestrial bodies ; but I answer that the experiments in terrestrial bodies come to no account ; for the attraction of homogeneous spheres near their surfaces are (by prop. 72, book I) as their diameters. Whence a sphere of one foot in diameter, and of a like nature to the earth, would attract a small body placed near its surface with a force 20,000,000 * times less than the earth would do if placed near its surface ; but so small a force could produce no sensible effect. If two such spheres were distant but by one-quarter of an inch, they would not, even in spaces void of resistance, come together by the force of their mutual attraction in less than a month's time ; †

* [*If the sphere is one foot in diameter, this number should be 40,000,000, since the diameter of the earth is about 40,000,000 ft. But perhaps Newton intended to say a sphere of one foot in radius*]

† [*The time is very much less. On the assumption that each of the spheres is one foot in diameter, Poynting (185, p 10) finds the time to be about 320.seconds If, however, we take one foot as the radius of each sphere, Todhunter (140, vol. 1, p 461) shows that the time is less than 250 seconds.*]

and less spheres will come together at a rate yet slower, viz., in the proportion of their diameters. Nay, whole mountains will not be sufficient to produce any sensible effect. A mountain of an hemispherical figure, three miles high and six broad, will not, by its attraction, draw the pendulum two minutes * out of the true perpendicular; and it is only in the great bodies of the planets that these forces are to be perceived, unless we may reason about smaller bodies in manner following.†

Let ABCD represent the globe of the earth cut by any plane, AC, into two parts, ACB and ACD. The part ACB bearing upon the part ACD presses it with its whole weight; nor can the part ACD sustain this pressure, and continue unmoved, if it is not opposed by an equal contrary pressure. And therefore the parts equally press each other by their weights—that is, equally attract each

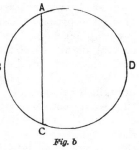

Fig. b

other, according to the third law of motion; and, if separated and let go, would fall towards each other with velocities reciprocally as the bodies. All which we may try and see in the loadstone, whose attracted part does not propel the part attracting, but is only stopped and sustained thereby.

Suppose now that ACB represents some small body on the earth's surface; then, because the mutual attractions of this particle, and of the remaining part ACD of the earth towards each other, are equal, but the attraction of the particle towards the earth (or its weight) is as the matter of the particle (as we have proved by the experiment of the pendulums), the attraction of the earth towards the particle will likewise be as the matter of the particle; and therefore the attractive forces of all terrestrial bodies will be as their several quantities of matter.

The forces (prop. 71, book I), which are as the matter in

* [*Maskelyne* (31) *says with reference to this :* " *It will appear, by a very easy calculation, that such a mountain would attract the plumb-line* 1' 18" *from the perpendicular.*"]

† [*This paragraph is of great importance, because in it Newton indicates the methods of all the experiments yet made in order to measure gravitational attraction in terrestrial bodies.*]

terrestrial bodies of all forms, and therefore are not mutable with the forms, must be found in all sorts of bodies whatsoever, celestial as well as terrestrial, and be in all proportional to their quantities of matter, because among all there is no difference of substance, but of modes and forms only. But in celestial bodies the same thing is likewise proved thus. We have shewn that the action of the circumsolar force upon all the planets (reduced to equal distances) is as the matter of the planets; that the action of the circumjovial force upon the satellites of Jupiter observes the same law; and the same thing is to be said of all the planets towards every planet; but thence it follows (by prop. 69, book I) that their attractive forces are as their several quantities of matter.

As the parts of the earth mutually attract one another, so do those of all the planets. If Jupiter and its satellites were brought together, and formed into one globe, without doubt they would continue mutually to attract one another as before. And, on the other hand, if the body of Jupiter was broken into more globes, to be sure, these would no less attract one another than they do the satellites now. From these attractions it is that the bodies of the earth and all the planets effect a spherical figure, and their parts cohere, and are not dispersed through the æther. But we have before proved that these forces arise from the universal nature of matter (prop. 72, book I), and that, therefore, the force of any whole globe is made up of the several forces of all its parts. And from thence it follows (by cor. 3, prop. 74) that the force of every particle decreases in the duplicate proportion of the distance from that particle; and (by prop. 73 and 75, book I) that the force of an entire globe, reckoning from the surface outwards, decreases in the duplicate, but, reckoning inwards, in the simple proportion of the distances from the centres, if the matter of the globe be uniform. And though the matter of the globe, reckoning from the centre towards the surface, is not uniform (prop. 73, book I), yet the decrease in the duplicate proportion of the distance outwards would (by prop. 76, book I) take place, provided that difformity is similar in places round about at equal distances from the centre. And two such globes will [1] (by the same proposition) attract one the other with a force decreasing in the duplicate proportion of the distance between their centres.

Wherefore the absolute force of every globe is as the quantity of matter which the globe contains ; but the motive force by which every globe is attracted towards another, and which, in terrestrial bodies, we commonly call their weight, is as the content under the quantities of matter in both globes applied to the square of the distance between their centres (by cor. 4, prop. 76, book I), to which force the quantity of motion, by which each globe in a given time will be carried towards the other, is proportional. And the accelerative force, by which every globe according to its quantity of matter is attracted towards another, is as the quantity of matter in that other globe applied to the square of the distance between the centres of the two (by cor. 2, prop. 76, book I); to which force the velocity by which the attracted globe will, in a given time, be carried towards the other is proportional. And from these principles well understood, it will be now easy to determine the motions of the celestial bodies among themselves.

SIR ISAAC NEWTON was born at Woolsthorpe, near Grantham, in Lincolnshire, in 1642. He was educated at the Grantham grammar-school, entered Trinity College, Cambridge, in 1661, and received his degree four years later. He at once began to make those magnificent discoveries in mathematics and physics which have made his name immortal. In 1665 he committed to writing his first discovery on fluxions, and shortly afterward made the unsuccessful attempt, to which we have already referred, to explain lunar and planetary motions. He next turned his attention to the subject of optics ; his work in that field includes the discovery of the unequal refrangibility of differently coloured lights, the compositeness of white light and chromatic aberration. Having erroneously concluded that this aberration could not be rectified by a combination of lenses, he turned his attention to reflectors for telescopes and made a great advance in that direction. His name is also closely identified with the colours due to thin plates. From 1669 to 1701 he was Lucasian professor of mathematics at Cambridge. He was elected to membership in the Royal Society in 1671, and from 1703 until his death was its president ; he became a member of the Paris Academy in 1699. The publication of his work on Optics had caused some controversy, and such a lover of peace was Newton, and so little did he care for the praise of the world, that it was only at the earnest solicitation of Halley that he was willing to give to the public the results of his wonderful researches on central orbits, and universal gravitation ; these included an explanation of the lunar inequalities, the figure of the earth, the precession of the equinoxes and the tides, and a method of comparing the masses of the heavenly bodies. In 1669 he became a member of Parliament, in 1696 Warden of the Mint, and from 1699 until his death was Master of the Mint. He gave much valuable aid in the recoinage of the money and in questions of finance at this period. He was knighted in 1705. During the latter years of his life much of his time was devoted to his public duties. He died in 1727, and was buried in Westminster Abbey.

20

LA FIGURE DE LA TERRE

Déterminée par les Observations de Messieurs Bouguer, et de la Condamine, de l'Académie Royale des Sçiences, envoyés par ordre du Roy au Pérou, pour observer aux environs de l'Équateur.

Avec une Relation abregée de ce Voyage, qui contient la description du Pays dans lequel les opérations ont été faites.

Par M. BOUGUER

À Paris, 1749

———

Section 7, pp. 327–394

———

THE FIGURE OF THE EARTH

Determined by the observations of MM. Bouguer and de la Condamine, of the Royal Academy of Sciences, sent to Peru by order of the King to make observations near the equator.

With a brief account of their travels and a description of the country in which the investigations were made.

By PIERRE BOUGUER

Paris, 1749

———

Section 7 pp. 327–394

CONTENTS OF SECTION VII

	PAGE
Introduction..	23
Chap. I.—Experiments Made in Order to find the Length of the Seconds-Pendulum	24
Description of Pendulum...............................	24
Method of Observation (Omitted).	
Observed Lengths of Seconds-Pendulum at Various Places	25
Corrections to be Made in the Observed Lengths	25
Corrected Lengths of Seconds-Pendulum at Various Places....	27
Chap. II.—Comparison of Attraction and Centrifugal Force (Omitted).	
Chap. III.—Remarks on the Diminution in Attraction at Different Heights above Sea-level..............................	27
Calculation of the Attraction Due to a Plateau...........	29
Deduction of the Mean Density of the Earth from Pendulum Experiments................................	32
Chap. IV.—On the Deflection of the Plumb-line by a Mountain..... ...	33
Description of Mount Chimborazo	34
Its Deflection of the Plumb-line Calculated from the Theory..	34
Various Ways Suggested for Showing the Deflection	35
Description of the Method Employed......................	38
Examination of the Attraction of Chimborazo	39
Meridian Altitudes at the First Station.....	40
Measurements Made to find the Relative Positions of the two Stations...	41
Meridian Altitudes at the Second Station (Omitted).	
Corrected Meridian Altitudes at the Second Station........	42
Calculations for the Observed Deflection of the Plumb-line...	42
Its Poor Agreement with that Calculated from Theory......	43
Appendix (Omitted).	

SECTION VII. OF BOUGUER'S FIGURE OF THE EARTH

ACCOUNT OF THE EXPERIMENTS OR OBSERVATIONS ON GRAV-
ITATION, WITH REMARKS ON THE CAUSES OF
THE FIGURE OF THE EARTH

1. HAVING discussed everything that bears on the earth considered as a geometrical body, it remains for us, before terminating this work, to verify the facts which give us some slight
knowledge of the interior conformation of this great mass considered as a physical body. . . .

2. The first question which presents itself on this matter is
a consideration of the part played in the flattening of the earth
by the attraction which compresses it from all sides, urging all
masses towards certain points. We know, since M. Richer first
remarked it (in 1672 in Cayenne), that this force is not every-
where the same. It is greater towards the poles, and less towards the equator. This agrees perfectly with the figure of the
earth, which appears to have yielded a little to the great pressure at the poles, and to be slightly elevated, on the contrary,
at the equator, where the compressing force was more feeble.
But does the effect correspond exactly to the cause upon which
we desire it to depend? Is the difference in attraction so great
that we can attribute to it all the inequality which exists, as
we have seen, between the two diameters of our globe? To
answer this question it is necessary to determine, by exact experiment, how much the attraction actually differs in different
parts of the earth. . . . We have two methods for observing
the change in attraction as we pass from one region to another;
we have only to examine how much more quickly or more
slowly a pendulum of given length oscillates; or else to find
the length of the pendulum whose time of vibration is exactly

a second ; the differences which we shall find in the length of this pendulum will determine the changes of the attraction as we go from one region to another.

I

ACCOUNT OF THE EXPERIMENTS MADE FOR THE PURPOSE OF DETERMINING THE LENGTH OF THE SECONDS-PENDULUM

3. My first experiments with the pendulum were made at Petit-Goave in the island of St. Domingue. They are reported in the memoirs of the Academy for 1735 and 1736. . . .

4. The instrument which I almost always used, and which I still use, is extremely simple. I make the pendulum always exactly of the same length, and I compare its oscillations with those of a clock which I regulate by daily observations. It is not, properly speaking, by the different lengths of the pendulum that I judge of the intensity of gravitation at different places ; I judge of it only by the greater or less rapidity of the oscillations, or by the number of oscillations made by the pendulum in 24 hours. . . . It appears to me to be much easier to count the number of oscillations than to measure directly differences of a few hundredths of a line* in the length of the pendulum.

[*Then follows an account of his pendulum. The bob was of copper, composed of two equal truncated cones joined at their greater bases. The thread was a fibre of aloe, which is not affected by the weather. The length was maintained constant by having it always so that an iron rule just fitted in between the clamp and the bob. The length of the equivalent simple pendulum was 36 pouces, 7.015 lines.*

Bouguer gives a description of a scale fixed behind the pendulum, by means of which he could observe the decrement and the time required by the pendulum to gain an oscillation on the clock.]

10. It is time to relate the experiments. . . . I shall choose one of those which I made on the rocky summit of Pichincha [2434 *toises above sea-level*], in the month of August, 1737. The

---

* [72 *pouces* = 1 *toise* = 1.949 *metres* = 6.3945 *ft.* 12 *lines* = 1 *pouce*]

force of attraction was feeble, not only because we were nearly over the equator at this place, but also because we were at a very great height above the surface of the sea. . . .

[*Details of experiment.*]

12. . . . We find in this way that the pendulum which beats seconds at the equator, and in the highest accessible place on the earth, is 36 pouces 6.69 lines in length. I made other experiments at the same place which agreed as exactly as possible with this result. [*One made by Don Antonio de Ulloa gave* 36 *pouces,* 6.715 *lines. We may take as the mean* 36 *pouces,* 6.70 *lines.*]

13. I have found by the same proceedings and with the aid of the same instruments, the length of the seconds-pendulum at Quito [1466 *toises above sea-level*], to be 36 pouces, 6.82 or 6.83 lines. I have verified it at different times and in all seasons of the year : at times of aphelion and perihelion, at the equinoxes, and when the sun was at intermediate points ; the extreme results were 36 pouces, 6.79 lines and 6.85 lines, with no differences which could not be attributed to the inevitable errors of observation. . . .

[*The question of a possible yearly change is discussed.*

Experiments were made with the same apparatus, in 1740, *at l'Isle de l'Inca,* 14' *or* 15' *from the equator, and scarcely* 40 *toises above sea-level. Bouguer regards this determination as that of the true equinoctial pendulum.*]

15.

Place	Length found by experiment
Under the equator at ⎰ 2434 toises absolute height.	36 pouces, 6 70 lines
⎱ 1466 " " "	" " 6 83 "
Sea-level	" " 7 07 "
At Portobello, 9° 34' N. latitude	" " 7 16 "
At Petit-Goave, 18° 27' " "	" " 7.33 "
At Paris	" " 8 58 "

CORRECTIONS WHICH MUST BE APPLIED TO THE LENGTH OF THE PENDULUM AS DETERMINED DIRECTLY FROM THE EXPERIMENTS.

16. [*Bouguer remarks that these corrections arise from changes in temperature and in the constitution of the atmosphere.*] The

first cause does not really change the length, it only makes it appear different according as the measures we use are differently altered by heat or cold; but the other cause brings in a real inequality, since it produces nearly the same effect as if the weight were greater or smaller. . . .

17. . . . Since the temperature of Quito does not differ from that of Paris in the middle of spring, we have only to refer all our results to it. That is, without altering the lengths of the pendulum found in these two cities, we have only to correct all the others by increasing or diminishing them, according as the metal rules we used were expanded by the heat or contracted by the cold. [*He concludes from his experiments that a change of length of pendulum of .02 lines corresponds to a change of temperature of 3° R. Hence he had to add .075 lines to the length found at sea-level, and subtract .05 lines from that found at Pichincha.*]

18. There is little more difficulty in finding the alteration in the length of the pendulum caused by the medium in which the experiments are made. This medium, whether rare or dense, has a certain weight, and that of the small mass of copper, of which the bob of the pendulum is formed, is a little lessened by it. The small mass tends to fall to the earth with only the excess of its weight above that of the air which surrounds it. Thus our pendulums are acted on by a force a little less than if we had performed the experiments *in vacuo:* and the length of the seconds-pendulum, which we found directly from experiment, is a little too short in the same proportion.

19. The use of the barometer enables us to find the ratio between the weight of mercury and of air in all the parts of the atmosphere which are accessible. We observe how many feet it is necessary to ascend or descend in order to change the height of the mercury by a line. . . . I have found in this way that it was only necessary to express the first (the weight of air) by unity, at the summit of Pichincha, if one expressed that of copper by 11000. . . . So I always found the seconds-pendulum too small by $\frac{1}{11000}$th part. To correct for this error we must add .04 lines [*at Pichincha ; .05 at Quito ; .06 at sea-level.*] . . . This is the first time that any one has taken account of this small correction which enters into the experiments, but we cannot neglect it if we wish to attain the greatest accuracy. . . .

THE LAWS OF GRAVITATION

[*Bouguer then proves that the time of vibration is not appreciably affected by the resistance of the air.*]*

22. Corrected lengths of the seconds-pendulum, or such as they would be if the oscillations were made *in vacuo*.

Place

	36 pouces, 6.69 lines
Under the equator at { 2434 toises absolute height.	36 pouces, 6.69 lines
1466 " "	" " 6.88 "
Sea-level	" " 7.21 "
At Portobello, 9° 34′ N. latitude	" " 7.30 "
At Petit-Goave, 18° 27′ " "	" " 7.47 "
At Paris	" " 8.67 "

II

COMPARISON OF ATTRACTION AND THE CENTRIFUGAL FORCE WHICH BODIES ACQUIRE BY THE MOTION OF THE EARTH ABOUT ITS AXIS, WITH REMARKS ON THE EFFECTS OF THESE TWO FORCES.

[*Bouguer finds that the primitive attraction (that is, the attraction the earth would have if it were at rest) is to the centrifugal force as* $288\frac{11}{30}$: 1. *He gives a table showing the decrease in the length of the seconds-pendulum at various latitudes, due to the centrifugal force. The following headings will give an idea of the matter contained in the rest of this chapter.*]

The centrifugal force produced by the motion of the earth about its axis is not sufficient to produce the observed differences in weight.

.   .   .   .

The primitive attraction does not tend towards a common point as centre.

.   .   .   .

III

REMARKS ON THE DIMINUTION IN THE ATTRACTION AT DIFFERENT HEIGHTS ABOVE THE LEVEL OF THE SEA.

40. The experiments with the pendulum which we have made at Quito and on the summit of Pichincha teach us that

* [*See note on page* 66]

the attraction changes with the distance from the centre of the earth. This force goes on diminishing as we ascend; I have found the pendulum at Quito to be shorter than at sea-level by .33 lines, or the $\frac{1}{1331}$th part: and in mounting to the summit of Pichincha the pendulum is shortened again by .19 lines, and is $\frac{1}{815}$th part shorter than at sea-level.* One cannot attribute these differences to the centrifugal force, which, being greater the higher we ascend, ought to diminish a little further the primitive attraction. The centrifugal force is increased by the height of the mountain by the $\frac{1}{1349}$th part only, and as it is itself but the $\frac{1}{289}$th part of the weight, it is clear that its new increase corresponds to .001 lines only in the length of the pendulum, and so does not sensibly contribute to the diminution of the other force.

41. If we compare the shortening which the pendulum receives with the height at which the experiment was made, we see that the forces do not decrease in the simple inverse ratio of the distances from the centre of the earth, but that they follow rather the proportion of the square. Quito is 1466 toises above sea-level, or $\frac{1}{2237}$th of the radius of the earth ; but it has been found that the attraction is less by a fraction much more considerable—namely, by a $\frac{1}{1331}$th part, which is nearly double ; this is not very far from the inverse ratio of the square of the distance. . . . We have a second example in the experiment made on Pichincha. The absolute height of this mountain, which is 2434 toises above sea-level, is $\frac{1}{1348}$th of the radius of the earth. The diminution of the length of the pendulum, or of the attraction, ought then to be the $\frac{1}{674}$th part, if it is to be in the inverse ratio of the square of the distance ; but it was by no means so great—in fact, only the $\frac{1}{815}$th part.

42. This diminution in attraction, as we go above sea-level, is quite in conformity with what we otherwise know. We can compare with the attraction here experimented upon that which keeps the moon in its orbit, or which obliges it continually to perform a circle about us. These two forces are exactly in the inverse ratio of the squares of the distances from the centre of the earth. We can make the same ex-

---

* [*Pendulum observations were made at these and other places in Peru by de la Condamine also* (8, pp 70, 144, 162–169) *For a complete bibliography of pendulum experiments, see that published by La Société Française de Physique* (178, vol 4)]

amination with respect to the principal planets which have several satellites, or with respect to the sun, towards which all the principal planets are attracted, and we shall always find the law of the square. Why, then, do our experiments constantly give a law not entirely in agreement with this? Is it necessary to attribute the difference to some error on our part ; or can it be that in the neighborhood of great masses like the earth the law under consideration is observed in an imperfect manner only ?

43. We shall find ourselves in a position to solve this difficulty, perhaps, by remarking that the Cordilleras, on which we were placed, form a kind of plateau, or, what in certain ways amounts to the same thing, the surface of the earth is there carried to a greater height or to a greater distance from the centre. There is reason for believing that in this second case the attraction would be a little greater ; for it is natural to think that it depends upon the size of the attracting mass. There are then two things to be considered in the case of the experiments on the pendulum which I have reported.

These experiments were made at a great height above the average surface of the earth, and therefore the attraction ought to be found a little less. But, on the other hand, the group of mountains on which Quito is placed and on which Pichincha rises, and all the other summits to which it acts as a plinth, ought to produce nearly the same effect as if the earth at this place were larger or had a greater radius. The attrac-

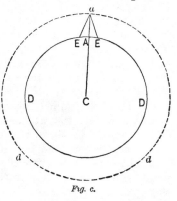

Fig. c.

tion on this account ought to increase. Thus it depends on a kind of chance, or, to speak more philosophically, it depends on circumstances which we do not yet know, whether the attraction at Quito will be equal to that at sea-level, or be smaller or larger.

44. Suppose that the circle ADD represents the circumference of the earth, of which C is the centre, and that Aa is the amount by which Quito, situated at a, is elevated

above sea-level. Imagine a new spherical shell of terrestrial matter, occupying all the interval between the two concentric surfaces ADD and *add;* or, which comes to the same thing, imagine that the earth increases in radius, and that Quito, without changing its position, remains at the level of the sea, now supposed much higher. There is every reason to think that the attraction at Quito would, as a consequence, be found greater than it actually is at A or at D, in the ratio of CA to C*a*. It is necessary for that, however, to suppose that the layer of earth enclosed between the two concentric surfaces is of the same density as all the rest; for if the density were different the increase would no longer be in the same ratio.

45. Call r the radius, and Δ the density of the earth. Then $r\Delta$ is the attraction at all the points A, D, etc., supposing that the earth ends there. Call h the height A*a*, which is very small compared with r. Then the attraction at a is less than at A, in the ratio of $r^2 : (r+h)^2$, or its diminution will be as $2h : r;$ that is, if the attraction is $r\Delta$ at A, it is $(r-2h)\Delta$ at a, and this supposes that the earth has CA only for effective radius. But all this will be subject to change if we add to our globe the layer A*d*D, whose density is δ. This new spherical layer, if it had the same density as the rest, would augment the attraction at the surface in the same ratio as the radius of the earth became greater. The increase would be in the ratio of $r : r+h$.

46. Thus the added layer would not only make up for the decrease which the attraction actually suffers when we go away from the earth, in rising by the height A*a*$=h$, but would add a new amount to it, equal to half the diminution, since it would make this attraction, which is actually $r-2h$ at the point a, become $r+h$. It follows that the attraction which the spherical layer can produce at its exterior surface at a is expressed by $3h$, or three times its thickness; but we must multiply by the density δ, because we suppose that the density of the layer and that of the earth as a whole are not equal.

47. To recapitulate: When the earth has its radius, CA$=r$, the attraction at A is $r\Delta$, and at the height h is $(r-2h)\Delta$. But when we add to the earth the spherical layer A*d*D, the attraction at a becomes $(r-2h)\Delta + 3h\delta$.

THE LAWS OF GRAVITATION

48. All that remains now to be remarked is that the Cordilleras of Peru, however great they may be, ought not to produce the same effect as the spherical shell which we have assumed. If the base EE of the Cordilleras were exactly double its height, and this mass had the shape of the roof of a house of indefinite length, then the Cordilleras would produce at a only $\frac{1}{4}$ the effect of the entire spherical shell, as can be easily proved. But there are further additions to be made in order to give a more accurate idea of the Cordilleras of Peru. The base EE is 80 or 100 times greater than the height Aa, which augments the effect in precisely the same ratio as the angle at a is greater. This angle is only 90° when we find the effect $\frac{1}{4}$ of that which the whole spherical layer would produce, but on account of the great width of the base of the Cordilleras the angle is nearer 170°, which doubles the effect. Moreover, the Cordilleras do not terminate at the height of Quito in a single summit like the ridge of a house ; it is, on the contrary, quite 10 or 12 leagues broad there. One can suppose then, without fear of mistake, that the effect is the greatest which can be produced by a chain of mountains. It is the $\frac{1}{2}$ of that which a spherical layer would produce, or $\frac{3}{2}h\delta$, and if we add to it the attraction $(r-2h)\Delta$, which the globe ADD produces at a, we shall have ($r-2h)\Delta + \frac{3}{2}h\delta*$ as the expression for the attraction at Quito, when $r\Delta$ expresses that at sea-level.

49. The difference between the two is $2h\Delta - \frac{3}{2}h\delta$, which furnishes the subject of divers quite curious remarks. If the matter of the Cordilleras were more compact than that of the average of the whole earth, and their densities were as 4 : 3, the difference $2h\Delta - \frac{3}{2}h\delta$ would become zero, and the attraction at Quito would be the same as at sea-level. If the density δ were still greater, our expression for the diminution would change sign and become an increase, so that the pendulum would be longer at Quito than at sea-level. But it is evident

---

* [*This formula is independently found by D'Alembert* (13, *vol.* 6, *pp.*85–92). *by Young* (51 *and* 95, *vol.* 2, *p.* 27), *and by Poisson* (65, *vol.* 1, *pp* 492–6) *Under the form* $g^1 = g_0 \left(1 - \dfrac{2h}{r} + \dfrac{3}{2}\dfrac{\delta h}{\Delta r}\right)$ *it is known as " Dr. Young'*⌐ *Rule," where* g^1 *is the value of gravity at height* h. *and* g_0 *is the value at the sea-level Faye* (147) *contends that the last term of the equation should be left out ; and if Airy's "flotation theory "* (94), *or Faye's compensation theory* (146$\frac{1}{2}$), *be true, there is no doubt that this term requires correction.*]

that things are not so. The difference in the length of the pendulum is sufficiently great to let us see that the density of the matter of which the Cordilleras is formed is much smaller than that of the rest of the globe.

50. We have found by experiment a diminution of a $\frac{1}{1331}$th part in the length of the pendulum, or in the attraction, as we go from the sea-level to Quito. So $\frac{1}{1331}$ corresponds to $2h\Delta - \frac{3}{2}h\delta$, as compared with $r\Delta$, which expresses the attraction at sea-level; that is, we have $\frac{1}{1331} = \frac{2h\Delta - \frac{3}{2}h\delta}{r\,\Delta}$. If we put $\frac{h}{r} = \frac{1}{2237}$, which is the ratio of the height of Quito to the radius of the earth, we shall have $\frac{1}{1331} = \frac{1}{2237} \times \frac{2\Delta - \frac{3}{2}\delta}{\Delta}$. Whence we deduce $\delta = \frac{850}{3993} \Delta$, which tells us that the Cordilleras of Peru, in spite of all the minerals they contain, have less than $\frac{1}{4}$ the density of the interior of the earth.*

51. We admit that this determination may contain a few errors on account of the large number of elements we had to employ in order to arrive at it. Nevertheless, if we once admit that the attraction, when the other circumstances are the same, follows exactly the direct ratio of the masses, we cannot doubt that the Cordilleras of Peru have a density considerably less than that of the rest of the globe. If we suppose Δ and δ equal, our expression for the difference of the attractions at Quito and at sea-level would become $\frac{h}{2r} \Delta$; which would make the difference between the lengths of the pendulum 4 times too small, or the attractions as the square roots, instead of the squares, of the distances from the centre of the earth. The attraction at Quito would be less than at sea-level by only the $\frac{1}{4474}$th part, and the pendulum would be really shorter by only 9 or 10 hundredths of a line, and in appearance by 2 or 3, on account of the different constitution and temperature of the air. The difference of the lengths of the pendulum is certainly greater. Thus it is necessary to admit that the earth is much more compact

* [*To give Bouguer's result more accurately, the density of the earth is* 4.7 *times that of the Cordilleras Saigey* (74, *p.* 149) *has made a recalculation of these results, with the proper reduction to vacuo, and finds* 4.25 *He has done the same for de la Condamine's pendulum experiments, with a result* 4.50. *For Addendum, see p* 160.]

below than above, and in the interior than at the surface. For the soil of Quito is like that of all other countries ; it is a mixture of earth and stones, with some metallic constituents. . . . Those physicists who imagined a great void in the middle of the earth, and who would have us walk on a kind of very thin crust, can think so no longer. We can make nearly the same objection to Woodward's theory of great masses of water in the interior. But let us continue to limit ourselves to the facts, or to the only immediate deductions which we can draw from them. These deductions are confirmed by the observations described in the next chapter, which is in the form I gave it in Peru before forwarding it to France.

IV

MEMOIR ON ATTRACTION AND ON THE MANNER OF OBSERVING WHETHER MOUNTAINS EXHIBIT IT (READ AT THE ACADÉMIE DES SCIENCES, IN OCTOBER, 1739)

52. It is very difficult not to accept attraction as a principle of fact or of experience. The most rigid Cartesians, like all other philosophers, cannot dispense with it in this sense. All they can do is to reserve to themselves the right of explaining it. . . . Since all the planets circle about the sun, there must necessarily be a force, I shall not say shoving them or drawing them, but rather transporting them at each moment towards this star. . . . Nothing prevents us from giving to this force the name " attraction," and from trying to assign to it a physical cause. . . .

[*Bouguer affirms that in establishing a new principle, it is not only necessary to prove the insufficiency of all others, but their impossibility also.*]

54. While waiting for all this to happen, it will contribute to the perfection of physics if we examine more carefully into attraction as a fact taught by experience. . . . It appeared to me that if all bodies act "at a distance," in proportion to their mass, and according to the other laws which we know, such enormous masses [*the mountains of Peru*] should produce a marked effect. I am well aware that they are very small compared with the whole earth ; but one can approach 1000 or 2000 times nearer their centre, and if it is true that

the attractions increase not only simply in the same ratio as the distances diminish, but in the inverse ratio of their squares, one ought to have a kind of compensation.

55. I shall content myself with justifying this in the case of a single mountain called Chimborazo, the base of which one is obliged to pass in going from the sea-side at Guayaquil to the more inhabited part of the province of Quito, which is enclosed between the two chains of mountains here formed by the Cordilleras, whose distance apart is 8 or 9 leagues. Chimborazo must be 3100 or 3200 toises above the sea-level [*he afterwards found it to be* 3217 *toises*], and 1700 or 1800 above the level of the plateau. We know exactly the relative heights of all the mountains we have seen, but not having yet been able to compare any one with sea-level, we are ignorant of their absolute heights. Chimborazo has roots which extend very far and become merged in those of the other mountains, so that it is very difficult to determine the true extent of its base. It must be more than 10,000 or 12,000 toises in diameter. But when we mount as high as possible, to where the snow begins, which is 850 toises from the top and renders the higher parts inaccessible, the mountain is still more than 3500 toises in diameter. The top, instead of terminating in a point, is rounded and blunt, and appears from below to have a width of 300 or 400 toises. From these dimensions one can estimate its huge mass. In the present investigation we need to know its height above ground only, not above sea-level. Even so it must be 20,000,000,000 cubic toises in volume. This is about the $\frac{1}{7,400,000,000}$th part only of the globe, and the effect of the attraction would be absolutely insensible, if one considered the quantity of matter only. But as we can place ourselves at 1700 or 1800 toises from the centre of gravity of the mountain, or 1900 times nearer to it than to the centre of the earth, this proximity ought to increase the effect about 3,600,000 times, and so make it about 2000 times less than that which gravitation produces, or the attraction caused by the whole mass of the earth. This we get by employing only a rough calculation and the lowest estimates. Calling the action of the mountain 1, and that of the earth 2000, the direction of attraction should be deflected from the vertical by about 1′ 43″. A plumb-line which would be directed exactly to the centre of

the earth, if its mass were exposed to the earth's attraction alone, ought then, on account of the action of the mountain, to be inclined by this same quantity, which is, as we see, quite considerable.

56. But how can we recognize this inclination; for all gravitating bodies must be equally subject to it, and we seem to lack a term of comparison? It would be useless to have recourse to the level surfaces of the heaviest liquids, since the attraction being equally altered with respect to them, their surfaces, instead of being perfectly horizontal, must suffer the same inclination. We see plainly, then, that, in order to judge of the amount of this alteration, it will be of no use to look just about us, we must seek another vertical line far off which is subject to no action from the mountain. But again, how are we to compare one vertical with another; or measure the angle which they make in meeting towards the centre of the earth, and that with sufficient accuracy? If while on the mountain, we observe with the quadrant the height of a point far off, and then go to that point and measure the height of the former place, it is true that by the difference of these two heights we can judge of the relative positions of the two vertical lines. But besides that we must know the exact distance from one to the other, it will be necessary also to suppose that the visual ray is a straight line; and it is not only certain that this is not true, we know that it is subject by refraction to a very irregular curvature. We cannot determine this curvature with sufficient exactness to enable us to find the effect of the attraction. It seems to me, therefore, that we must seek in the heavens a term of comparison. By this means, however, we shall easily overcome every difficulty; and what a moment ago seemed an impossibility becomes at once very simple.

57. We have but to station ourselves to the north or to the south of a mountain, and as near as possible to its centre of gravity, and observe the latitude. This observation can be made with the greatest accuracy only by using a quadrant or other equivalent instrument whose plumb-line will be deflected toward the mountain; this is the same as saying that the zenith will recede from the mountain. Then we must go east or west of this station to such a distance that the attraction is negligible; and if we observe the latitude in this

second place with the same care and with the same means as in the first, it is evident that all the difference which we shall observe will be due to attraction. In order to have this second station precisely east or west of the first, we must observe the azimuth of the sun at its rising or setting, by finding its position with reference to some easily distinguished point on the horizon; in doing so we must often suppose the latitude known; but the error we may make on this supposition will be of no consequence, and it will always be easy to find two stations on the same parallel of latitude to within 3 or 4 sixtieths of a second. The latitude will be found precisely the same in the two places, if the vertical line has not been altered in the first. Suppose, however, that without seeking the latitude, we observe simply the meridian altitudes of a star at the two stations; the difference of these two altitudes will indicate equally well the deflection of the vertical line. It is evident that all the stars which pass the meridian on the side of the apparent vertical line next to the mountain will appear lower at the first station than at the second; for as the plumb-line approaches the mountain the apparent zenith recedes from it and from these stars. It will be quite the reverse with those stars which pass the meridian on the other side of the apparent vertical line: they will appear higher at the first station.*

58. Instead of taking the stations both to the north or both to the south, we could take them one to the north and the other to the south, and exactly on the same meridian; then the effect of the attraction would be doubled, roughly speaking, and we should find the sum of the contrary attractions. The vertical line would be inclined in opposite directions at the two stations; and the altitudes of stars which would be increased in the one would be decreased in the other. The physical effect being doubled would be more sensible, and more susceptible of observation. If the two points were equally distant from the centre of gravity of the mountain, the action would be equal at both, and in order to get each

* [*This method of doubling the deflection caused by the mountain, by observing not one star, but at least two, one north and one south of the stations, is due to de la Condamine. See his account of the expedition* (8, *p.* 68), *Zach* (49) *and Poynting* (185, *p.* 14). *This is the method actually employed by Bouguer.*]

of them we should have merely to take half of the quantity
furnished by the comparison of the observations. In other
cases the division would be a little more difficult; nevertheless it would be sufficient, as we shall shew later, to divide
the sum of the contrary attractions proportionally to the pro-
ducts of the quantity by which each station is more north or
more south, respectively, than the centre of gravity of the
mountain and the cube of the distance of the other station,
respectively, from the same centre. Thus we are under the
necessity of knowing the situation of each station with reference to the mountain; but we must know the distance from
one station to the other also, in order to determine geometric-
ally the difference of latitude between them. It is evident
that this difference must itself produce a change in the altitude of each star, and we must know it before we can tell what
is the double effect of the attraction. To obtain the difference
in latitude of the two places, it would suffice ordinarily to
measure to the east or to the west of the mountain a base
directed nearly north and south, and to form on this base two
triangles which end at the two stations.

59. This way of making two observations from different
sides of the same mountain in order to render the effect of the
attraction more sensible, seems to me the more useful method,
as it depends less on the peculiarities of the places. We can
sometimes double the effect also by making the first observation at the north of one mountain and the second at the
south of another. If the two stations are not exactly on the
same east and west line, we have only to determine geometric-
ally their difference of latitude, and take account of it in the
comparison of the altitudes of the stars.

60. Finally, it is not only by observations made at the north
or at the south that we can discover whether mountains are
capable of acting "at a distance"; it can be done also by observations made at the east or the west; but with this difference, that it will be no longer a question of observing latitude,
or of taking the meridian altitudes of stars; it will be only a
question of determining time exactly. It appears to me that
this last method would be often preferable to the preceding
ones, except that it requires two observers. Suppose that the
first of these is on the east side of a mountain, and the second
on the west side of another, or of the same, mountain. If each

of them regulates carefully a chronometer by corresponding altitudes, it is evident that all these altitudes being altered by the attraction which deflects the plumb-line, each chronometer will be regulated as if the meridian were not exactly vertical, but inclined below toward the mountain, and above away from it.   Let us suppose that the attraction amounts to a minute of arc, and that the two mountains are on the equator ; the first chronometer will denote midday 4 seconds of time too soon, and the other 4 seconds too late.   Thus, neglecting the difference of longitude, which we could easily find by measuring trigonometrically the distance of the two observers apart and reducing this distance to degrees and minutes, there would be a difference of 8 seconds of time between the two chronometers.   If the two mountains instead of being on the equator were at latitude 60°, each minute of inclination which the attraction produced in a plumb-line would produce 8 seconds of difference in the time of midday, and therefore 16 seconds difference in the chronometers. Finally, to judge of the attraction we need only know the exact difference between the chronometers ; and to find this, it would always be sufficient to agree upon a signal, by fire or otherwise ; and to observe at both stations the minute and second of the instantaneous appearance of this signal.

61. I return to the first method because it appears to me to be the simplest : that is, suppose we station ourselves always to the north or to the south of the mountain and confine ourselves to observations of the latitude.   It is evident that if we take at each station the meridian altitude of one star only, we must know to the last degree of nicety the condition of the quadrant we are using.   There is no lack of methods for verifying this instrument, but there is one which is extremely valuable in the present instance, because, at the same time as we work at verifying the quadrant, we are making the observations which decide the question at issue ; and in thus abridging the operations we avoid opportunities for errors.   This method is to take the meridian altitudes of an equal number of stars toward the north and toward the south, and, provided that the state of the instrument does not vary from one observation to another, it does not matter if it does change from day to day.   If it makes the altitudes of the stars on one side the zenith too great, it will produce the same effect with re-

spect to those on the other side. Thus the change will influence only the sum of the altitudes or the complements of the altitudes, and will not alter the difference of the altitudes taken on the different sides. The attraction, on the contrary, will not alter the sum, but will change the difference; because at the same time that it makes the stars on one side too high, it makes those on the other side too low. It will always be easy to separate these two causes, and we shall not attribute to the one that which arises from the other. To obtain at one stroke the effect of attraction without being obliged to know the state of the quadrant or the declinations of the stars, we need only examine whether the differences of the meridian altitudes taken towards the north and towards the south are the same at the two stations, or whether they are subject to a second difference. But it is necessary to remark that the altitudes being increased on the one side while they are diminished on the other, it is the half of this second difference which denotes the physical effect of the attraction, both when this effect is single and when it is double. In this latter case, it will be necessary to divide the total effect in the ratio which the separate effects ought to have.

[*Bouguer then proves this ratio to be that mentioned above (p. 37). He admits that some mountains might shew less attraction than that required by Newton's law (or even none), due to the existence of great cavities in the mass. He discusses the different mountains in the neighbourhood of Quito, and for various reasons decides upon Chimborazo as the one most suitable for the experiment.*]

EXAMINATION OF THE ATTRACTION OF CHIMBORAZO

65. I did not ascend this mountain alone as I did the preceding one. I had some time before communicated my design and all my views to M. de la Condamine, and when on the point of carrying them out I mentioned them to M. de Ulloa, one of the two naval lieutenants who had assisted in the observations both of myself and of M. de la Condamine ever since our arrival in the domains of His Catholic Majesty. These gentlemen obligingly offered to accompany me, not only in the preparatory examination, but also during the stay it was necessary to make on the mountain side ; and as I knew it would be to the advantage of the observations, I hastened to accept the

offer. I had already thought that Chimborazo fulfilled approximately the necessary conditions : I knew that it was very easy of access; it could be seen from Quito, or rather from Pichincha, from which it was 75,000 toises distant; and I had already measured its height. . . . On December 4th we established ourselves on the south side of the mountain, at the bottom of the snow line, 829 toises below the summit, but about 2400 above sea-level, and exactly 910 toises above the place at Quito where I have always made my observations, and 344 toises above that part of Pichincha where there is a cross which can be seen from all parts of the city, and where I passed some days in March, 1737, in order to observe the astronomical refraction. I shall not speak of the cold and the other discomforts we had to put up with ; snow covered our tent and all the ground around as far as 800 or 900 toises below us, and we lived in fear of being buried under its weight. It needed continual vigilance in order to avoid it. [*M. de Ulloa fell ill, and had to descend the mountain on December 15th.*]

[*Left alone, Bouguer and de la Condamine observed the altitudes of 10 stars, 4 on the south side and 6 on the north. The following are the altitudes as affected by the error of the instrument and by refraction; they are the means of the readings of the two observers.*]

67.

	Meridian altitudes at the first station					
	On 14th Dec			On 15th Dec		
On the north side :	o	′	″	o	′	″
Capella	42	50	42½	42	50	30
First head of Gemini.				56	6	5
Second head of Gemini.	59	53	55	59	53	57¼
Second horn of Aries	66	18	15	66	18	50
First horn of Aries	69	0	20	69	0	17¼
Aldebaran	72	33	42½	72	33	57½
On the south side :						
Acarnar	32	58	10	32	58	25
Canopus	38	58	52½	38	58	55
Tail of Cetus	72	6	36½	72	6	35
Sirius	75	9	12½	75	9	40

68. We observed the meridian altitude of the sun three times. De la Condamine found it at the lower edge on December 15th to be 67° 54′ 26″. This altitude, which is corrected for the error of the instrument, but not for refraction

and parallax, gives 1° 29′ 53″ south for the latitude of the place where we were. I observed it on the 5th and 12th; on the 5th we had not yet regulated the chronometer nor traced the meridian, and I found 1° 30′ 16″. On the 12th I observed the apparent altitude of the lower edge of the sun to be 68° 5′ 34″; which gives 1° 30′ 6″.

69. As soon as we were established on Chimborazo, I had sent a tent about a league and a half to the west to a place called *l'Arénal* to serve as the second station. [*Bouguer then describes the measurements made to find the exact position of the second station; it was* 3570 *toises distant from the first,* 174 *toises lower, and somewhat south of west of it. They began their observations from the second station on Dec.* 16th. *Here they suffered more from the wind and cold than at the more elevated station, as they were more exposed to the prevailing east wind. It filled their eyes with dust and continually threatened to overturn the tents.* * *The screws of the quadrant could not be turned at night without applying heat to them. Then follows a table of the altitudes as observed. Since the second station was* 505 *toises, or* 32″, *south of the first, we must increase by* 32″ *all the altitudes of stars observed toward the north, and diminish the others by the same amount. Moreover, since the second station was lower than the first by* 174 *toises, the altitudes observed at the second station must be diminished, to reduce them to the level of the first, on account of the excess of astronomical refraction. The following are the corrected altitudes, and the differences for each star of the mean determinations at the two stations :*]

---

* [*For de la Condamine's account of the experiments, see his Journal* (8, *p.* 69 *and* 8½, *pt.* 2, *p.* 146).]

	CORRECTED ALTITUDES AT THE SECOND STATION						Excess of altitudes at the first station over those at the second, after the latter have been corrected
	On 21st Dec			On 22d Dec			
	°	′	″	°	′	″	′ ″
On the north side:							
Capella.................	42	49	10	42	49	15	1 24
First head of Gemini...		—			—		—
Second head of Gemini.		—		59	53	23½	—
Second horn of Aries...	66	16	59	66	17	16	1 25
First horn of Aries.....	68	59	6	68	59	11	1 10
Aldebaran.............	72	32	31½	72	32	36½	1 6*
On the south side:							
Acarnar......	32	56	53	32	56	28	1 37½
Canopus...............	38	57	16	38	57	36	1 28
Tail of Cetus..........	72	4	47½	72	4	50	1 48
Sirius.................	75	8	0	75	8	7½	1 22½

[*Bouguer considers the observations on the tail of Cetus and the first horn of Aries as the best, but thinks it most legitimate to take the mean of the altitudes of each star at each station and to give equal weight to their differences. These differences are given in the last column of the preceding table, and are, he maintains, too large and too uniform to be due to any defect in the observations. The averages of the excess of all the stars on the north side and all those on the south side are now to be taken.*]

74. ... They give about 1′ 19″ as the mean excess for the north stars, and 1′ 34″ for the south. The second difference is 15″. I leave it to my readers to say whether such a quantity is sufficiently established by the means employed. My quadrant was 2.5 ft. in radius, and it must be remarked that any errors which may exist in its graduation are of no importance here; since we have to do not with the altitudes themselves, but with their differences. Suppose we admit the 15″, it will give 7″.5 for the effect of attraction; it would be much greater if we compared the tail of Cetus with the first horn of Aries. However this is not the complete and absolute effect; for if attraction really takes place, the mountain must have some effect at the second station, which was about 4572 toises from the centre of the mountain, and 61°.5 to the west of south. At the first station we were nearly 16° west of south, and 1753

* [*This is evidently a misprint for* 1′ 16″.]

toises distant. From these data we find that the effect at the
nearer station is to the effect we ought to find at the other as
$1358 : 100$, or as $13\frac{7}{12} : 1$ nearly. But since our observations
give only the difference of the two effects, we must increase
$7''.5$ by a 13th or 14th part of itself in order to have the total
effect [*which makes it $8''$*].

75. We must admit that this effect is very different from
what we had expected. But we know so little about the
earth's density, and on the other hand that of the mountain
may be so different from that which we have assumed it to be,
that there is no reason to be surprised at anything.* [*It is a
tradition among the natives that Chimborazo is an extinct volcano, and, if so, its density would be very hard to estimate.
Bouguer thinks it might be better to experiment on smaller and
denser mountains.*] It is very probable that we shall find in
France or in England some hill of sufficient size, especially if
we double the effect; and I shall be delighted if I find on my
return that the experiments that shall have been made either
confirm mine or throw new light on the matter. [At Riobamba
in Peru, December 30, 1738.]

[*In an appendix Bouguer states that after a more thorough
survey of the Cordilleras he failed to find a more satisfactory
place at which to repeat his experiment. He suggests that the
converse effect be experimented upon : viz., the decrease in gravity due to some deep cañon among mountains. Assuming such
great cavities in Chimborazo as would make its real only half
its apparent volume, he finds his results would make it 6 or
7 times† less dense than the earth ; this he thinks not unreason-*]

* [*De la Condamine also laid little stress on the numerical result of the plumb-
line experiment ; for he says* (8, p. 69), " *if we can deduce from it nothing
decisively in favour of the Newtonian attraction, at least we find nothing contrary to that theory* "]

† [*In a critical analysis of all the experiments made to determine the density
of the earth up to that time, Saigey* (74, p. 151 *and in his* "*Petite Physique du
Globe,*" *Paris*, 1842, *pt.* 2, *p.* 151), *in* 1842, *stated that Bouguer's calculations
were erroneous, because he confused the centre of attraction with the centre of
gravity of the mountain ; he refers to a method, by means of which, using
Bouguer's own mean result, he deduced the density of the earth to be* 4.62
*times that of the mountain ; but adds that if he had used only the results from
observations on the tail of Cetus and the first horn of Aries, the result would
have been* 1.83, *which is almost exactly the same as that found by Maskelyne by
the same method for the hill Schehallien.*]

able. Bouguer further remarks that the distribution of density in the earth may be such that the maximum attraction of the earth is not at its surface but at some distance beneath it.]

[*In connection with this work of Bouguer should be read a paper* (9) *presented by him to the Academy on April* 28, 1756, *on the possibility of detecting the deflection of a plumb-line due to the ebb and flow of the tide.*

Valuable accounts and discussions of Bouguer's work in Peru are given by von Zach (43, 44, *and* 49), *Schmidt* (64, *vol.* 2, *p.* 475), *Todhunter* (140. *chap.* 12), *Zanotti-Bianco* (148½, *pt.* 2, *pp.* 122–25), *and Poynting* (185, *pp.* 10–14.]

PIERRE BOUGUER was born in 1698 at Croisic, Bretagne, and was educated at the Jesuit College at Vannes. He succeeded his father as professor of hydrography at Croisic in 1713, and in 1730 accepted a similar position at Havre. His investigations concerning the intensity of light, embodied in a work, *Essai d'optique sur la gradation de la lumière,* published in 1729, led to his being elected a member of the Academy of Sciences in 1731; he was promoted to the office of pensioned astronomer in 1735. Along with two other members of the Academy, MM. de la Condamine and Godin, he was sent to Peru in 1735 to measure the length of an arc of the meridian near the equator. Their labors there lasted ten years, and the results of their observations were published in 1749 in *La Figure de la Terre,* from which we have given the preceding extracts. For several years afterwards Bouguer was engaged in a bitter controversy with de la Condamine concerning their respective shares in the Peruvian researches. In addition to his works on photometry, he published several valuable treatises on navigation, and various papers on atmospheric refraction and other optical problems, and on mechanics. He died at Paris in 1758.

THE BERTIER CONTROVERSY

THE BERTIER CONTROVERSY

In June, 1769, there appeared in the *Journal des Sciences et des Beaux Arts* a letter (11) by a M. Coultaud, who signed himself Former Professor of Physics at Turin. In it he described some pendulum experiments made in the Alps of Savoy. He claimed to have found that at a height of 1085 toises above the base of the mountain the pendulum gained 28′ in 2 months; 20′ 22″ in 3 months at a height of 514 toises; and 15′ 4″ in 175 days at a height of 210 toises. So that it appeared as if the attraction of gravitation increased with the distance from the earth's centre, instead of behaving according to the Newtonian law. A full account of the apparatus and observations was added, and there seemed no reason why credence should not be given to the results. The advocates of the Newtonian theory felt called upon to account for this phenomenon consistently with their doctrine. D'Alembert (12 and 13, vol. 6, pp. 85–92) attacked the problem and found that the Newtonian theory was adequate to explain the fact, provided the mean density of the earth were about three eighths of that of the mountain.* An abstract of Coultaud's alleged observations is given by David (14).

In Dec., 1771, another letter appeared in the same journal (15) signed by one Mercier, and addressed to Gessner, Professor of Physics in the Univ. of Geneva. It described experiments made in Valois similar to those of Coultaud and with similar results, namely, that the attraction of gravitation is *directly* proportional to the square of the distance. D'Alembert then discussed the question again (13, vol. 6, pp. 93–98). Further explanations on the Newtonian theory were forth-

---

* Compare the conclusion of Bouguer on p. 31 of this volume.

coming from Le Sage (16) and Lalande (17). Roiffé also discussed the experiments (18).

These results of Coultaud and Mercier seem, however, to have been a cause of great exultation to a certain number of scientists, especially ecclesiastics, who contended that the Newtonians wished to take from them their Father, their God, in asserting that bodies attract and move of themselves without any Prime Mover. It is hard to believe that this feeling existed so late as a century ago. One of the most active of the opponents of the theory of Newton was Father Bertier, de l'Oratoire, who founded his 4th volume of *Les Principes Physiques* on the above experiments. For several years a warm discussion raged among French physicists over the question.

Le Sage, having had his suspicions aroused by some passage in Mercier's letter, began a careful investigation into the genuineness of the experiments both of Coultaud and Mercier. He found them to be fabrications from beginning to end (19). Le Sage does not mention whom he supposes to be the perpetrators or instigators of the fraud.

A new impetus was given to the discussion by the publication of 2 letters (20) from Father Bertier describing experiments with the balance similar to those performed by members of the Royal Society of London a century before ; but Bertier writes as if the idea were entirely a new one. The length of the string used to suspend the weight from one arm of the balance, after it had been counterpoised in the pan above, was 74 ft. In one case weights of 25 lbs. were used, and when one of them was suspended at the end of the string it lost in weight 1 ounce 3.5 drachms. Bertier concluded, much to his satisfaction, that bodies weigh more the farther they are from the centre of the earth. Roiffé followed with a paper (21) discussing the experiments made thus far and remarking that Bertier had not taken account of the difference in the density of the air at the two levels. Le Sage also criticised Bertier very harshly (22). Repetitions of Bertier's experiment were made by M. David, and Fathers Cotte and Bertier (23), the one with a string of 170 ft. and weights of 1220 lbs., the others with a string of 45 ft. and weights of 150 lbs. The one reported a loss in weight of 1 oz., the others of 2 lbs., in the same direction as indicated by Bertier's first experiment. An article by David (24) in answer to Le Sage contains some scornful strict-

ures on Newton and his principles which form amusing reading at this late date. Rozier (25) criticised all the experiments made on the laws of gravitation ; he refers to some
more made by Bertier (26) from which the latter concluded
that the loss in weight was proportional to the length of string
and to the weight. Rozier then announced the details of some
experiments of a similar kind made by himself, which gave
quite discordant results. David wrote another letter (27) with
more details of his experiments, but adding nothing of value.
Bertier followed with a similar letter (28). A committee of
the Academy of Dijon repeated the experiments with a sensitive balance, and found (29) no change in weight except that
due to the different densities of the air at the higher and
lower levels. Some experiments on this same subject were
made by Achard (33) ; he found by using first a string and
then a brass chain with which to suspend the masses, that the
changes in weight of the suspended mass could be ascribed to
variations in the temperature and dampness of the air. Dolomieu (30) made some experiments with a weight suspended in
a mine, similar to those of Dr. Power (page 2) ; his results
permitted no definite conclusions as to change in weight. In
connection with this work we might notice a valuable article
by Le Sage (34) on the history of the theory of gravitation and
the experiments made concerning it. He gives a brief account
of the views of Gilbert, Bacon, Kepler, Beaugrand, Fermat,
Pascal, Roberval, Descartes, and Gassendi ; finally he discusses Dolomieu's experiments and the possible variation in
density beneath the surface of the earth.

The controversy closes with the appearance of a letter (36)
from Bertier making an humble retraction of his statements
regarding the deductions to be drawn from his experiment ; he
admits that such experiments do not prove that bodies weigh
more as they are farther from the earth, but he declines to give
up his belief that such is the case. He again inveighs against
those who " by means of 101 different laws, which they make
God create to cover their ignorance, explain everything with a
facility that is truly delightful."

THE SCHEHALLIEN EXPERIMENT

51

THE SCHEHALLIEN EXPERIMENT

In 1772, Maskelyne, the English Astronomer Royal, proposed to the Royal Society (31) that the experiment of Bouguer on the attraction of a mountain be repeated in Great Britain, as Bouguer himself had suggested 30 years before. Maskelyne had been informed of two places which might be convenient for the purpose. One was near the confines of Yorkshire and Lancashire, on the hill Whernside; the other in Cumberland, on the hill Helvellyn. The proposal was favourably received by the Society, and Mr. Charles Mason was sent to examine various hills in England and Scotland, and to select the most suitable (32). Mason found that the two hills referred to by Maskelyne were not suitable; and fixed upon Schehallien in Perthshire as offering the best situation. At the earnest solicitation of the Royal Society, Maskelyne himself undertook to make the necessary observations. He had at his disposal a 10-foot zenith sector, and all his other instruments were the best of their kind at the time. The work was begun in the summer of 1774. The method of finding the deflection of the plumb-line due to the hill was exactly the second of the methods described by Bouguer (page 36); he took readings of the zenith-distances of certain stars at two stations, one north and one south of the hill, and by this means doubled the deflection of the plumb-line. Between June 30th and September 22d he took 169 star observations from the south station, and 168 from the north station; in all 337 observations on 43 stars. At the same time a very elaborate survey by triangulation was made of the dimensions and form of the hill. This was considered as made up of a very large number of prisms, sufficient data for the determination of each of which were collected during the survey.

In his paper (32) describing the operations, Maskelyne calculates, from 40 only out of the 337 observations, that the apparent difference of latitude between the two stations is 54".6.* The true difference of latitude is 43", leaving 11".6 due to the contrary attractions of the hill.

From a rough calculation, assuming the density of the mountain to be the same as the mean density of the earth, and that the law of attraction is that of the inverse square of the distance, Maskelyne found that the attraction should be twice that found by observation. Hence the mean density of the earth is twice that of the hill. A more exact calculation was promised for the future. Maskelyne draws two main conclusions: (1), that Schehallien has an attraction, and so, therefore, has every mountain; (2), that the inverse square law of the distance is confirmed; for if the force were only a little affected by the distance, the attraction of the hill would be wholly insensible.

The survey of the hill and its environs was made during the years 1774, 1775 and 1776. The calculation of the attraction of the hill from these measurements was undertaken by Hutton,† who employed several new and interesting methods. A full account will be found in his paper (37 and 47, vol. 2, pp. 1–68). Assuming that the density of the hill is the same as the mean density of the earth, Hutton found that the attraction of the earth is to the sum of the contrary attractions of the hill as 9933 : 1. Now Maskelyne had found the deflection due to the contrary attractions of the hill to be 11".6; whence the attraction of the earth is to the sum of the attractions of the hill as 1 : tan. 11".6, or as 17781 : 1; or, allowing for the

---

* Von Zach (49, App., pp 686–692) has calculated the results from all of the 337 observations, and finds for the apparent difference of latitude 54".651, and for the deflection due to the contrary attractions of the hill 11".632; which is in entire accord with Maskelyne's calculations. Saigey (74, p 153) also subjected the result to a test which was satisfactory. Zanotti-Bianco states (148½, pt. 2, p. 134) that Saigey maintained that Maskelyne did not choose his station at the most favourable part of the hill-side, and that if he had done so he would have found the deflection 14" instead of 11".6.

† For Hutton's own estimate of his share in the work, and for his contempt for Cavendish's experiment, see bibl. No. 45. For a good account of Hutton's method of calculation, see Zanotti-Bianco (148½, pt. 2, pp. 126–32); see also Helmert (148, vol 2, pp. 368–80).

centrifugal force, as 17804 : 1 nearly. Hence the mean density of the earth is to the density of the hill as 17804 : 9933, or as 9 : 5 nearly. Assuming the specific gravity of the hill to be about 2.5, Hutton remarks that this would give 4.5 as the mean specific gravity of the earth. Hutton revised this result in his "Tracts" (47, vol. 2, p. 64); he takes the specific gravity of the hill as 3, and hence the specific gravity of the earth would be 5.4 nearly.

Playfair, with the aid of Lord Webb Seymour, made a careful lithological survey of Schehallien, and published his results in 1811 (46 and 48). He found that the hill was made up of two classes of rock, quartz of specific gravity 2.639876, and micaceous rock, including calcareous, of specific gravity 2.81039. From two suppositions as to the distribution of these two components in the interior of the hill, using Hutton's data for the attraction, Playfair calculated the mean density of the earth to be 4.55886 and 4.866997 respectively. Playfair considered the experiment on Schehallien so exact that he took the mean of the above results, 4.713, as the best determination of the mean density of the earth.

Hutton prefers to take 2.77 as the mean of Playfair's determinations for the density of the hill, and the density of the earth as $\frac{9}{5}$ of 2.77, or 5 nearly. In a paper published in 1821 (52, 53 and 54), Hutton complains that his share in the Schehallien experiment has always been underestimated ; he gives a brief account of the observations, calculations and results, and considers 5 as the most probable value of the mean density of the earth. He shews that the Schehallien experiment could not be made to give the same result as that of Cavendish, 5.48, unless the deflection 11″.6 be diminished to about 10″.5 or 10″.4, which is manifestly too great an error to have been committed by Maskelyne, considering the accuracy of the observer and of the instruments, and the large number of observations made. Hutton suggests the repetition of the experiment at one of the pyramids in Egypt. Some years later Peters (80½) made a calculation of the attraction of the Great Pyramid.

For brief accounts of the Schehallien experiment and criticisms upon it, reference should be made to Hutton (38 and 47, vol. 2, pp. 69–77), von Zach (43, 44 and 49), Muncke (61, vol. 3, pp. 944–70), Schmidt (64, vol. 2, pp. 474–9), Menabrea (71),

Schell (135), Todhunter (140, vol. 1, pp. 459–69). Zanotti-Bianco (148½, pt. 2, pp. 125–35) and Fresdorf (186½, pp. 5–7).

Capt. Jacob has remarked (118 and 121) that by this method we may measure the attraction of the mass of the mountain *above* the surface, yet we do not know how much ought to be added or subtracted due to that *below* it.

Von Zach makes mention of several early astronomers who assign anomalies in their geodetic measurements to the influence of mountains on the plumb-lines of their instruments; the reader is referred to von Zach, Humboldt (82, vol. 1, notes, pp. 45–7) and Helmert (148, vol. 2, chap. 4), and to the account in this volume (p. 123) of the work of James and Clarke. Von Zach himself made a very careful determination in 1810, after the method used by Bouguer, of the attraction of mount Mimet, near Marseilles. He found a deflection of the plumb-line amounting to 2″. He did not calculate the density of the earth. His observations were published in book form in 1814 (49).

For this work Maskelyne was presented by the Royal Society with the Copley medal. At the presentation the President, Sir John Pringle, delivered an address (35) on the attraction of gravitation, giving a critical account of the state of the subject before the time of Newton, as well as of its later developments.

EXPERIMENTS TO DETERMINE THE DENSITY OF THE EARTH

BY

HENRY CAVENDISH, Esq., F.R.S. and A.S.

Read June 21, 1798

(From the Philosophical Transactions of the Royal Society of London for the year 1798, *Part II.*, pp. 469–526)

CONTENTS

	PAGE
Introduction	59
Description of the apparatus	61
Method of observing the deflection	64
" " " " *time of vibration*	64
Effect of the resistance of the air	65
Account of the experiments....	67
Testing for magnetic effects	68
Testing the elastic properties of the wire	72
Further tests for magnetic effects	75
Testing the effect of variation of temperature about the box	76
Final observations....	80
On the theory of the experiment	88
Corrections to be made in the theory as first given	91
Effect of the variable position of the arm on the equations	97
When and how to apply the corrections	98
Table of results	99
Conclusion	99
Appendix : to find the attraction of the mahogany case on the balls	102

EXPERIMENTS TO DETERMINE THE DENSITY OF THE EARTH

HENRY CAVENDISH, Esq., F.R.S. and A.S.

MANY years ago, the late Rev. John Michell, of this society, contrived a method of determining the density of the earth, by rendering sensible the attraction of small quantities of matter ; but, as he was engaged in other pursuits, he did not complete the apparatus till a short time before his death, and did not live to make any experiments with it.   After his death, the apparatus came to the Rev. Francis John Hyde Wollaston, Jacksonian Professor at Cambridge, who, not having conveniences for making experiments with it, in the manner he could wish, was so good as to give it to me.

The apparatus is very simple ; it consists of a wooden arm, 6 feet long, made so as to unite great strength with little weight. This arm is suspended in an horizontal position, by a slender wire 40 inches long, and to each extremity is hung a leaden ball, about 2 inches in diameter ; and the whole is inclosed in a narrow wooden case, to defend it from the wind.

As no more force is required to make this arm turn round on its centre, than what is necessary to twist the suspending wire, it is plain, that if the wire is sufficiently slender, the most minute force, such as the attraction of a leaden weight a few inches in diameter, will be sufficient to draw the arm sensibly aside. The weights which Mr. Michell intended to use were 8 inches diameter.   One of these was to be placed on one side the case, opposite to one of the balls, and as near it as could conveniently be done, and the other on the other side, opposite to the other ball, so that the attraction of both these weights would conspire in drawing the arm aside ; and, when its position, as affected by these weights, was ascertained, the weights were to be removed to the other side of the case, so as to draw the arm the contrary way, and the position of the arm was to be again determined ; and, consequently, half the difference of these positions would shew how much the arm was drawn aside by the attraction of the weights.

In order to determine from hence the density of the earth, it is necessary to ascertain what force is required to draw the arm aside through a given space. This Mr. Michell intended to do, by putting the arm in motion, and observing the time of its vibrations, from which it may easily be computed.*

Mr. Michell had prepared two wooden stands, on which the leaden weights were to be supported, and pushed forwards, till they came almost in contact with the case ; but he seems to have intended to move them by hand.

As the force with which the balls are attracted by these weights is excessively minute, not more than $\frac{1}{50,000,000}$ of their weight, it is plain, that a very minute disturbing force will be sufficient to destroy the success of the experiment; and, from the following experiments it will appear, that the disturbing force most difficult to guard against, is that arising from the variations of heat and cold ; for, if one side of the case is warmer than the other, the air in contact with it will be rarefied, and, in consequence, will ascend, while that on the other side will descend, and produce a current which will draw the arm sensibly aside.†

* Mr. Coulomb has, in a variety of cases, used a contrivance of this kind for trying small attractions ; but Mr. Michell informed me of his intention of making this experiment, and of the method he intended to use, before the publication of any of Mr. Coulomb's experiments.

† M. Cassini, in observing the variation compass placed by him in the observatory (which was constructed so as to make very minute changes of position visible, and in which the needle was suspended by a silk thread), found that standing near the box, in order to observe, drew the needle sensibly aside ; which I have no doubt was caused by this current of air It must be observed, that his compass-box was of metal, which transmits heat faster than wood, and also was many inches deep ; both which causes served to increase the current of air. To diminish the effect of this current, it is by all means advisable to make the box, in which the needle plays, not much deeper than is necessary to prevent the needle from striking against the top and bottom.

THE LAWS OF GRAVITATION

As I was convinced of the necessity of guarding against this source of error, I resolved to place the apparatus in a room which should remain constantly shut, and to observe the motion of the arm from without, by means of a telescope ; and to suspend the leaden weights in such manner, that I could move them without entering into the room. This difference in the manner of observing, rendered it necessary to make some alteration in Mr. Michell's apparatus ; and, as there were some parts of it which I thought not so convenient as could be wished, I chose to make the greatest part of it afresh.

Fig. 1 is a longitudinal vertical section through the instrument, and the building in which it is placed. ABCDDCB-AEFFE is the case ; x and x are the two balls, which are suspended by the wires hx from the arm $ghmh$, which is itself suspended by the slender wire gl. This arm consists of a slender deal rod hmh, strengthened by a silver wire hgh ; by which means it is made strong enough to support the balls, though very light.*

The case is supported, and set horizontal, by four screws, resting on posts fixed firmly into the ground ; two of them are represented in the figure, by S and S ; the two others are not represented, to avoid confusion. GG and GG are the end walls of the building. W and W are the leaden weights ; which are suspended by the copper rods $RrPrR$, and the wooden bar rr, from the centre pin Pp. This pin passes through a hole in the beam HH, perpendicularly over the centre of the instrument, and turns round in it, being prevented from falling by the plate p. MM is a pulley, fastened to this pin ; and Mm, a cord wound round the pulley, and passing through the end wall ; by which the observer may turn it round, and thereby move the weights from one situation to the other.

Fig. 2 is a plan of the instrument. AAAA is the case. SSSS, the four screws for supporting it. hh, the arm and balls. W and W, the weights. MM, the pulley for moving them. When

---

* Mr. Michell's rod was entirely of wood, and was much stronger and stiffer than this, though not much heavier ; but, as it had warped when it came to me, I chose to make another, and preferred this form, partly as being easier to construct and meeting with less resistance from the air, and partly because, from its being of a less complicated form, I could more easily compute how much it was attracted by the weights.

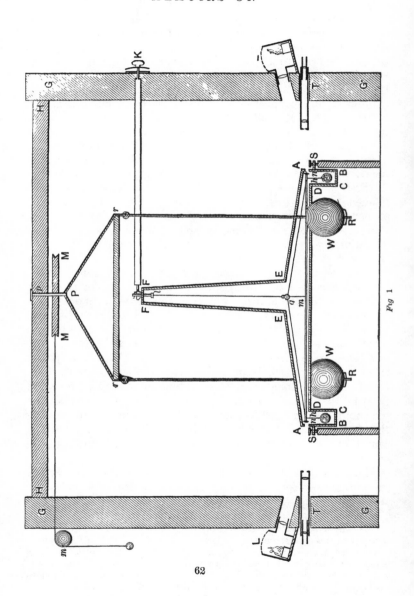

Fig. 1

the weights are in this position, both conspire in drawing the arm in the direction hW ; but, when they are removed to the situation w and w, represented by the dotted lines, both conspire in drawing the arm in the contrary direction $h w$. These weights are prevented from striking the instrument, by pieces of wood, which stop them as soon as they come within $\frac{1}{8}$ of an inch of the case. The pieces of wood are fastened to the wall

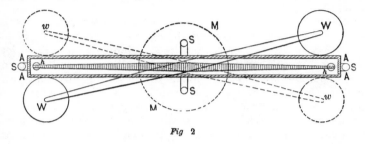

Fig 2

of the building ; and I find, that the weights may strike against them with considerable force, without sensibly shaking the instrument.

In order to determine the situation of the arm, slips of ivory are placed within the case, as near to each end of the arm as can be done without danger of touching it, and are divided to 20ths of an inch. Another small slip of ivory is placed at each end of the arm, serving as a vernier, and subdividing these divisions into 5 parts ; so that the position of the arm may be observed with ease to 100ths of an inch, and may be estimated to less. These divisions are viewed, by means of the short telescopes T and T (Fig. 1), through slits cut in the end of the case, and stopped with glass ; they are enlightened by the lamps L and L, with convex glasses, placed so as to throw the light on the divisions ; no other light being admitted into the room.

The divisions on the slips of ivory run in the direction Ww (Fig. 2), so that, when the weights are placed in the positions w and w, represented by the dotted circles, the arm is drawn aside, in such direction as to make the index point to a higher number on the slips of ivory ; for which reason, I call this the positive position of the weights.

FK (Fig. 1) is a wooden rod, which, by means of an endless screw, turns round the support to which the wire gl is fastened, and thereby enables the observer to turn round the wire, till the arm settles in the middle of the case, without danger of touching either side. The wire *gl* is fastened to its support at top, and to the centre of the arm at bottom, by brass clips, in which it is pinched by screws.

In these two figures, the different parts are drawn nearly in the proper proportion to each other, and on a scale of one to thirteen.

Before I proceed to the account of the experiments, it will be proper to say something of the manner of observing. Suppose the arm to be at rest, and its position to be observed, let the weights be then moved, the arm will not only be drawn aside thereby, but it will be made to vibrate, and its vibrations will continue a great while ; so that, in order to determine how much the arm is drawn aside, it is necessary to observe the extreme points of the vibrations, and from thence to determine the point which it would rest at if its motion was destroyed, or the point of rest, as I shall call it. To do this, I observe three successive extreme points of a vibration, and take the mean between the first and third of these points, as the extreme point of vibration in one direction, and then assume the mean between this and the second extreme, as the point of rest ; for, as the vibrations are continually diminishing, it is evident, that the mean between two extreme points will not give the true point of rest.

It may be thought more exact, to observe many extreme points of vibration, so as to find the point of rest by different sets of three extremes, and to take the mean result ; but it must be observed, that notwithstanding the pains taken to prevent any disturbing force, the arm will seldom remain perfectly at rest for an hour together ; for which reason, it is best to determine the point of rest, from observations made as soon after the motion of the weights as possible.

The next thing to be determined is the time of vibration, which I find in this manner : I observe the two extreme points of a vibration, and also the times at which the arm arrives at two given divisions between these extremes, taking care, as well as I can guess, that these divisions shall be on different sides of the middle point and not very far from it. I then compute the middle point of the vibration, and, by proportion, find the time at which the arm comes to this middle point. I then, after a

number of vibrations, repeat this operation, and divide the interval of time, between the coming of the arm to these two
middle points, by the number of vibrations, which gives the
time of one vibration. The following example will explain
what is here said more clearly:

Extreme points	Divisions	Time	Point of rest	Time of middle of vibration
27.2				
	25 24	10^h 23′ 4″ ⎰ 57 ⎱	—	10^h 23′ 23″
22 1	—	—	24 6	
27.	—	—	24 7	
22.6	—	—	24 75	
26 8	—	—	24.8	
23.	—	—	24.85	
26.6	—	—	24.9	
	25 24	11 5 22 ⎰ 6 48 ⎱	—	11 5 22
23.4				

The first column contains the extreme points of the vibrations. The second, the intermediate divisions. The third,
the time at which the arm came to these divisions; and the
fourth, the point of rest, which is thus found: the mean between the first and third extreme points is 27.1, and the mean
between this and the second extreme point is 24.6, which is
the point of rest, as found by the three first extremes. In like
manner, the point of rest found by the second, third, and
fourth extremes, is 24.7, and so on. The fifth column is the
time at which the arm came to the middle point of the vibration, which is thus found: the mean between 27.2 and 22.1 is
24.65, and is the middle point of the first vibration; and, as
the arm came to 25 at 10^h 23′ 4″, and to 24 at 10^h 23′ 57″, we
find, by proportion, that it came to 24.65 at 10^h 23′ 23″. In
like manner, the arm came to the middle of the seventh vibration at 11^h 5′ 22″; and, therefore, six vibrations were performed
in 41′ 59″, or one vibration in 7′ 0″.

To judge of the propriety of this method, we must consider
in what manner the vibration is affected by the resistance of the
air, and by the motion of the point of rest.

Let the arm, during the first vibration, move from D to B
(Fig. 3), and, during the second, from B to d; Bd being less
than DB, on account of the resistance. Bisect DB in M, and

E   65

Bd in m, and bisect Mm in n, and let x be any point in the vibration; then, if the resistance is proportional to the square of the velocity, the whole time of a vibration is very little altered; but, if T is taken to the time of one vibration, as the diameter of a circle to its semi-circumference, the time of moving from B to n exceeds $\frac{1}{2}$ a vibration, by $\dfrac{T \times Dd}{8Bn}$ nearly; and the time of moving from B to m falls short of $\frac{1}{2}$ a vibration,

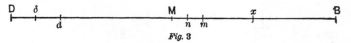

Fig. 3

by as much; and the time of moving from B to x, in the second vibration, exceeds that of moving from x to B, in the first, by $\dfrac{T \times Dd \times Bx^2}{4Bn^2 \times \sqrt{Bx} \times x\delta}$, supposing D$d$ to be bisected in δ; so that, if a mean is taken, between the time of the first arrival of the arm at x and its returning back to the same point, this mean will be earlier[*] than the true time of its coming to B, by $\dfrac{T \times Dd \times Bx^2}{8Bn^2 \sqrt{Bx} \times x\delta}$.

The effect of motion in the point of rest is, that when the arm is moving in the same direction as the point of rest, the time of moving from one extreme point of vibration to the other is increased, and it is diminished when they are moving in contrary directions; but, if the point of rest moves uniformly, the time of moving from one extreme to the middle point of the vibration, will be equal to that of moving from the middle point to the other extreme, and, moreover, the time of two successive vibrations will be very little altered; and, therefore, the time of moving from the middle point of one vibration to the middle point of the next, will also be very little altered.

---

[*] [*This word should be "later," as is observed by Todhunter* (140, *vol* 2, *p.* 165) *For an elementary discussion of this kind of motion see Williamson and Tarleton's " Treatise of Dynamics," ex.* 13, § 117 *Poisson* (65, *vol.* 1, *pp.* 353–361) *and Menabrea* (71) *have given very elaborate analyses of the problem. Cornu and Baille* (137, 141, 142, 143, *and* 157) *proved in* 1878 *that the resistance in the case under consideration is proportional to the first power of the velocity.*]

It appears, therefore, that on account of the resistance of
the air, the time at which the arm comes to the middle point
of the vibration, is not exactly the mean between the times of
its coming to the extreme points, which causes some inaccur-
acy in my method of finding the time of a vibration. It must
be observed, however, that as the time of coming to the middle
point is before the middle of the vibration, both in the first
and last vibration, and in general is nearly equally so, the error
produced from this cause must be inconsiderable ; and, on the
whole, I see no method of finding the time of a vibration which
is liable to less objection.

The time of a vibration may be determined, either by previous
trials, or it may be done at each experiment, by ascertaining the
time of the vibrations which the arm is actually put into by
the motion of the weights ; but there is one advantage in the
latter method, namely, that if there should be any accidental
attraction, such as electricity, in the glass plates through which
the motion of the arm is seen, which should increase the force
necessary to draw the arm aside, it would also diminish the
time of vibration ; and, consequently, the error in the result
would be much less, when the force required to draw the arm
aside was deduced from experiments made at the time, than
when it was taken from previous experiments.

ACCOUNT OF THE EXPERIMENTS

In my first experiments, the wire by which the arm was suspended was $39\frac{1}{4}$ inches long, and was of copper silvered, one
foot of which weighed $2\frac{4}{10}$ grains ; its stiffness was such as to
make the arm perform a vibration in about 15 minutes. I immediately found, indeed, that it was not stiff enough, as the
attraction of the weights drew the balls so much aside, as to
make them touch the sides of the case ; I, however, chose to
make some experiments with it, before I changed it.

In this trial, the rods by which the leaden weights were suspended were of iron ; for, as I had taken care that there should
be nothing magnetical in the arm, it seemed of no signification
whether the rods were magnetical or not ; but, for greater se-
curity, I took off the leaden weights, and tried what effect the
rods would have by themselves. Now I find, by computation,
that the attraction of gravity of these rods on the balls, is to

that of the weights, nearly as 17 to 2500 ; so that, as the attraction of the weights appeared, by the foregoing trial, to be sufficient to draw the arm aside by about 15 divisions, the attraction of the rods alone should draw it aside about $\frac{1}{10}$ of a division ; and, therefore, the motion of the rods from one near position to the other, should move it about $\frac{1}{5}$ of a division.

The result of the experiment was, that for the first 15 minutes after the rods were removed from one near position to the other, very little motion was produced in the arm, and hardly more than ought to be produced by the action of gravity ; but the motion then increased, so that, in about a quarter or half an hour more, it was found to have moved $\frac{1}{2}$ or $1\frac{1}{2}$ division, in the same direction that it ought to have done by the action of gravity. On returning the irons back to their former position, the arm moved backward, in the same manner that it before moved forward.

It must be observed, that the motion of the arm, in these experiments, was hardly more than would sometimes take place without any apparent cause ; but yet, as in three experiments which were made with these rods, the motion was constantly of the same kind, though differing in quantity from $\frac{1}{2}$ to $1\frac{1}{2}$ division, there seems great reason to think that it was produced by the rods.

As this effect seemed to me to be owing to magnetism, though it was not such as I should have expected from that cause, I changed the iron rods for copper, and tried them as before ; the result was, that there still seemed to be some effect of the same kind, but more irregular, so that I attributed it to some accidental cause, and therefore hung on the leaden weights, and proceeded with the experiments.

It must be observed, that the effect which seemed to be produced by moving the iron rods from one near position to the other, was, at a medium, not more than one division ; whereas the effect produced by moving the weight from the midway to the near position, was about 15 divisions ; so that, if I had continued to use the iron rods, the error in the result caused thereby, could hardly have exceeded $\frac{1}{30}$ of the whole.

THE LAWS OF GRAVITATION

EXPERIMENT I.   AUG. 5

Weights in midway position

Extreme points	Divisions	Time			Point of rest	Time of mid of vibration	Difference
	11.4	9^h 42'	0''				
	11.5	55	0				
	11.5	10 5	0		11.5		

At 10^h 5', weights moved to positive position

Extreme points	Divisions	Time			Point of rest	Time of mid of vibration	Difference
23.4							
27.6	—	—			25.82		
24.7	—	—			26.07		
27.3	—	—			26.1		
25.1	—	—					

At 11^h 6', weights returned back to midway position

Extreme points	Divisions	Time			Point of rest	Time of mid of vibration	Difference
5.							
	11	0 0	48	⎱	—	0^h 1' 13''	
	12	1	30	⎰			
18.2	—	—			12	—	14' 56''
	12	16	29	⎱	—	16 9	
	11	17	20	⎰			
6.6	—	—			11.92	—	14 36
	11	30	24	⎱	—	30 45	
	12	31	11	⎰			
16.3	—	—			11.72	—	15 13
	12	45	58	⎱	—	45 58	
	11	47	4	⎰			
7.7							

Motion on moving from midway to pos = 14.32

pos. to midway = 14.1

Time of one vibration = 14' 55''

It must be observed, that in this experiment, the attraction of the weights drew the arm from 11.5 to 25.8, so that, if no contrivance had been used to prevent it, the momentum acquired thereby would have carried it to near 40, and would, therefore, have made the balls to strike against the case. To prevent this, after the arm had moved near 15 divisions, I returned the weights to the midway position, and let them remain there, till the arm came nearly to the extent of its vibration, and then again moved them to the positive position, whereby the vibrations were so much diminished, that the balls did not touch the sides; and it was this which prevented my observing the first extremity of the vibration. A like method was used, when the weights were returned to the midway position, and in the two following experiments.

The vibrations, in moving the weights from the midway to the positive position, were so small, that it was thought not worth while to observe the time of the vibration. When the weights were returned to the midway position, I determined the time of the arm's coming to the middle point of each vibration, in order to see how nearly the times of the different vibrations agreed together. In great part of the following experiments, I contented myself with observing the time of its coming to the middle point of only the first and last vibration.

Experiment II. Aug. 6

Weights in midway position

Extreme points	Divisions	Time	Point of rest	Time of mid of vibration	Difference
	11	10ʰ 4′ 0″			
	11	11 0			
	11	17 0			
	11	25 0	11.		

Weights moved to positive position

29.3					
24.1	—	—	26.87		
30.	-	—	27.57		
26 2	—	—	28.02		
29.7	—	—	28.12		
26.9	—	—	28.05		
28.7	—	—	27.85		
27.1	—	—	27.82		
28.4					

Weights returned to midway position

6.					
	12	1 3 50 ⎫	—	1ʰ 4′ 1″	
	13	4 34 ⎭			
18.5	—	—	12.37	—	14′ 52″
	13	18 29 ⎫	—	18 53	
	12	19 18 ⎭			
6.5	—	—	11.67	—	14 46
	11	33 48 ⎫	—	33 39	
	12	34 51 ⎭			
15.2	—	—	11.	—	13 46
	13	45 8 ⎫	—	47 25	
	12	46 22 ⎭			
7.1	—	—	10 75	—	15 25
	11	2 3 48 ⎫	—	2 2 50	
	12	5 18 ⎭			
13 6					

Motion of arm on moving weights from midway to pos = 15 87
pos. to midway = 15.45

Time of one vibration = 14′ 42″

Experiment III. Aug. 7

The weights being in the positive position, and the arm a little in motion

Extreme points	Divisions	Time	Point of rest	Time of mid of vibration	Difference
31.5					
29	—	—	30.12		
31	—	—	30.02		
29.1					
Weights moved to midway position					
9.					
	14	10ʰ 34′ 18″ ⎫	—	10ʰ 34′ 55″	
	15	35 8 ⎭			
20.5	—	—	14.8	—	14′ 44″
	15	49 31 ⎫	—	49 39	
	14	50 27 ⎭			
9.2	—	—	14.07	—	14 38
	14	11 5 7 ⎫	—	11 4 17	
	15	6 18 ⎭			
17.4	—	—	13.52	—	14 47
	14	18 46 ⎫	—	19 4	
	13	19 58 ⎭			
10 1	—	—	13.3	—	14 27
	13	33 46 ⎫	—	33 31	
	14	35 26 ⎭			
15.6					
Weights moved to positive position					
32.					
	28	0 2 48 ⎫	—	0 2 59	
	27	3 56 ⎭			
23.7	—	—	27.8		
31.8	—	—	28.27		
25.8	—	—	28.62		
	27	44 58 ⎫	—	47 40	
	28	46 50 ⎭			
31.1					

Motion of the arm on moving weights from pos. to mid. = 15.22

mid. to pos. = 14.5

Time of one vibration, when in mid position = 14′ 39″

pos. position = 14′ 54″

These experiments are sufficient to shew, that the attraction of the weights on the balls is very sensible, and are also sufficiently regular to determine the quantity of this attraction pretty nearly, as the extreme results do not differ from each other by more than $\frac{1}{10}$ part. But there is a circumstance in them, the reason of which does not readily appear, namely, that the effect of the attraction seems to increase, for half an hour, or an hour, after the motion of the weights ; as it may be observed, that in all three experiments, the mean position kept increasing for that time, after moving the weights to the positive position ; and kept decreasing, after moving them from the positive to the midway position.

The first cause which occurred to me was, that possibly there might be a want of elasticity, either in the suspending wire, or something it was fastened to, which might make it yield more to a given pressure, after a long continuance of that pressure, than it did at first.

To put this to the trial, I moved the index so much, that the arm, if not prevented by the sides of the case, would have stood at about 50 divisions, so that, as it could not move farther than to 35 divisions, it was kept in a position 15 divisions distant from that which it would naturally have assumed from the stiffness of the wire ; or, in other words, the wire was twisted 15 divisions. After having remained two or three hours in this position, the index was moved back, so as to leave the arm at liberty to assume its natural position.

It must be observed, that if a wire is twisted only a little more than its elasticity admits of, then, instead of setting, as it is called, or acquiring a permanent twist all at once, it sets gradually, and, when it is left at liberty, it gradually loses part of that set which it acquired ; so that if, in this experiment, the wire, by having been kept twisted for two or three hours, had gradually yielded to this pressure, or had begun to set, it would gradually restore itself, when left at liberty, and the point of rest would gradually move backwards ; but, though the experiment was twice repeated, I could not perceive any such effect.

The arm was next suspended by a stiffer wire.

EXPERIMENT IV.   AUG. 12

Weights in midway position

Extreme points	Divisions	Time	Point of rest	Time of mid of vibration	Difference
	21.6	9ʰ 30′ 0″			
	21.5	52 0			
	21.5	10 13 0	21.5		

Weights moved from midway to positive position

27.2					
22 1	—	—	24.6		
27.	—	—	24.67		
22 6	—	—	24 75		
26 8	—	—	24 8		
23 0	—	—	24.85		
26 6	—	—	24 9		
23.4					

Weights moved to negative position

15.						
	17	19 25 ⎱	—	10ʰ 20′ 31″		
	19	20 41 ⎰				
22.4	—	—	18.72	—		7′ 0″
	20	26 45 ⎱	—	27 31		
	19	27 22 ⎰				
15.1	—	—	18.52	—		6 57
	19	35 1 ⎱	—	34 28		
	20	48 ⎰				
21.5	—	—	18 35	—		7 23
	20	40 23 ⎱	—	41 51		
	19	41 18 ⎰				
15.3	—	—	18.22	—		6 48
	18	48 36 ⎱	—	48 39		
	19	49 24 ⎰				
20.8	—	—	18 1	—		6 58
	19	54 45 ⎱	—	55 37		
	18	55 45 ⎰				
15.5						

Weights moved to positive position

31.3						
	25	11 10 25 ⎱	—	11 10 40		
	23	11 3 ⎰				
17.1	—	—	24.02	—		7 3
	22	17 6 ⎱	—	17 43		
	23	26 ⎰				
30.6	—	—	24.17	—		7 1
	25	24 33 ⎱	—	24 44		
	23	25 17 ⎰				
18.4	—	—	24 32	—		7 5
	23	31 21 ⎱	—	31 49		
	25	32 9 ⎰				
29 9	—	—	24 4	—		6 59
	25	38 39 ⎱	—	38 48		
	23	39 31 ⎰				
19.4	—	—	24 5	—		7 6
	23	45 16 ⎱	—	45 54		
	25	46 12 ⎰				
29.3						

Motion of arm on moving weights from midway to pos. = 3.1

pos. to neg. = 6.18

neg. to pos. = 5.92

Time of one vibration in neg. position = 7′ 1″

pos. position = 7′ 3″

Experiment V.   Aug. 20

The weights being in the positive position, the arm was made to vibrate, by moving the index

Extreme points	Divisions	Time			Point of rest	Time of mid of vibration			Difference
29.6									
21.1	—		—		25.2				
29.	—		—		25.17				
21.6									
Weights moved to negative position									
22.6									
	20	10ʰ 22′	47″	}	—	10ʰ 23′	11″		
	19	23	30	}					
16.3	—		—		19.27				
21.9	—		—		19.15				
16.5	—		—		19.1				
21.5	—		—		19.07				
16.8	—		—		19.07				
21.2	—		—		19.07				
17.1	—		—		19.05				
20.8	—		—		19.02				
17.4	—		—		19.05				
20.6	—		—		19.02				
	20	11 32	16	}	—	11 33	53		
	19	33	58	}					
17.5	—		—		18.97	—			7′ 13″
	19	41	16	}	—	41	6		
	20	43	0	}					
20.3									
Weights moved to positive position									
20.2									
	24	49	10	}	—	49	37		
	26	50	19	}					
29.4	—		—		24.95	—			7 7
	26	56	15	}	—	56	44		
	25		47	}					
20.8	—		—		24 92				
28.7	—		—		24.87				
21.3	—		—		24.85				
28.1	—		—		24.75				
21.5	—		—		24.67				
27.6	—		—		24.67				
22.	—		—		24.7				
	24	0 45	48	}	—	0 46	21		
	25	46	43	}					
27.2	—		—		24.7	—			7 1
	25	53	11	}	—	53	22		
	24	54	9	}					
22.4									

Motion of arm on moving weights from pos. to neg. $= 5.9$
neg. to pos. $= 5.98$
Time of one vibration, when weights are in neg. position $= 7′\ 5″$
pos. position $= 7′\ 5″$

In the fourth experiment, the effect of the weights seemed to increase on standing, in all three motions of the weights, conformably to what was observed with the former wire ; but in the last experiment the case was different ; for though, on moving the weights from positive to negative, the effect seemed to increase on standing, yet, on moving them from negative to positive, it diminished.

My next trials were, to see whether this effect was owing to magnetism. Now, as it happened, the case in which the arm was inclosed, was placed nearly parallel to the magnetic east and west, and therefore, if there was anything magnetic in the balls and weights, the balls would acquire polarity from the earth ; and the weights also, after having remained some time, either in the positive or negative position, would acquire polarity in the same direction, and would attract the balls ; but, when the weights were moved to the contrary position, that pole which before pointed to the north, would point to the south, and would repel the ball it was approached to ; but yet, as repelling one ball towards the south has the same effect on the arm as attracting the other towards the north, this would have no effect on the position of the arm. After some time, however, the poles of the weight would be reversed, and would begin to attract the balls, and would therefore produce the same kind of effect as was actually observed.

To try whether this was the case, I detached the weights from the upper part of the copper rods by which they were suspended, but still retained the lower joint, namely, that which passed through them ; I then fixed them in their positive position, in such manner, that they could turn round on this joint, as a vertical axis. I also made an apparatus by which I could turn them half way round, on these vertical axes, without opening the door of the room.

Having suffered the apparatus to remain in this manner for a day, I next morning observed the arm, and, having found it to be stationary, turned the weights half way round on their axes, but could not perceive any motion in the arm. Having suffered the weights to remain in this position for about an hour, I turned them back into their former position, but without its having any effect on the arm. This experiment was repeated on two other days, with the same result.

We may be sure, therefore, that the effect in question could not be produced by magnetism in the weights; for, if it was, turning them half round on their axes, would immediately have changed their magnetic attraction into repulsion, and have produced a motion in the arm.

As a further proof of this, I took off the leaden weights, and in their room placed two 10-inch magnets; the apparatus for turning them round being left as it was, and the magnets being placed horizontal, and pointing to the balls, and with their north poles turned to the north; but I could not find that any alteration was produced in the place of the arm, by turning them half round; which not only confirms the deduction drawn from the former experiment, but also seems to shew, that in the experiments with the iron rods, the effect produced could not be owing to magnetism.

The next thing which suggested itself to me was, that possibly the effect might be owing to a difference of temperature between the weights and the case; for it is evident, that if the weights were much warmer than the case, they would warm that side which was next to them, and produce a current of air, which would make the balls approach nearer to the weights. Though I thought it not likely that there should be sufficient difference, between the heat of the weights and case, to have any sensible effect, and though it seemed improbable that, in all the foregoing experiments, the weights should happen to be warmer than the case, I resolved to examine into it, and for this purpose removed the apparatus used in the last experiments, and supported the weights by the copper rods, as before; and, having placed them in the midway position, I put a lamp under each, and placed a thermometer with its ball close to the outside of the case, near that part which one of the weights approached to in its positive position, and in such manner that I could distinguish the divisions by the telescope. Having done this, I shut the door, and some time after moved the weights to the positive position. At first, the arm was drawn aside only in its usual manner; but, in half an hour, the effect was so much increased, that the arm was drawn 14 divisions aside, instead of about three, as it would otherwise have been, and the thermometer was raised near 1°.5; namely, from 61° to 62°.5. On opening the door, the weights were found to be no more heated, than just to prevent their feeling cool to my fingers.

As the effect of a difference of temperature appeared to be so great, I bored a small hole in one of the weights, about three-quarters of an inch deep, and inserted the ball of a small thermometer, and then covered up the opening with cement. Another small thermometer was placed with its ball close to the case, and as near to that part to which the weight was approached as could be done with safety; the thermometers being so placed, that when the weights were in the negative position, both could be seen through one of the telescopes, by means of light reflected from a concave mirror.

Experiment VI. Sept. 6

Weights in midway position

Extreme points	Divisions	Time	Point of rest	Thermometer	
				in air	in weight
	18.9	9ʰ 43′	—	55.5	
	18.85	10 3	18.85	—	
Weights moved to negative position					
13.1	—	10 12	—	55.5	55.8
18.4	—	18	15.82		
13.4	—	25			
missed					
13 6	—	39	—	55.5	55.8
17.6	—	46	15.65		
13.8	—	53	15.65		
17.4	—	11 0	15.65		
14.0	—	7	15.65		
17.2	—	14	—	55.5	
Weights moved to positive position					
25.8	—	23			
17.5	—	30	21.55		
25.4	—	37	21 6		
18.1	—	44	21 65		
25.0	—	51			
missed					
24.7	—	0 5			
19.	—	12	21.77		
24.4	—	19			

Motion of arm on moving weights from midway to − = 3.03
− to + = 5.9

Experiment VII. Sept. 18

Weights in midway position

Extreme points	Divisions	Time	Point of rest	Thermometer	
				in air	in weight
	19.4	8ʰ 30′	—	56.7	
	19.4	9 32	—	56.6	
Weights moved to negative position					
13.6	—	40	—	—	57.2
18.8	—	47	16.25		
13.8	—	54			
Eight extreme points missed					
16.9	—	10 58			
14.5	—	11 5	15.62		
16.6	—	12			
Weights moved to positive position					
26.4	—	20	—	56.5	
17.2	—	28	21.72		
26.1	—	35			
Four extreme points missed					
19.3	—	0 10			
25.1	—	17	22.3		
19.7	—	24			

Motion of arm on moving weights from midway to — = 3.15
— to + = 6.1

Experiment VIII. Sept. 23

Weights in midway position

Extreme points	Divisions	Time	Point of rest	Thermometer	
				in air	in weight
	19.3	9ʰ 46′	—	53.1	
	19.2	10 45	19.2	53.1	
Weights moved to negative position					
13.5	—	56	—	—	53.6
18.6	—	11 3	16.07		
13.6	—	10			
Four extreme points missed					
17.4	—	44			
14.1	—	51	15.7		
17.2	—	58	—	—	53.6

Weights moved to positive position

15.7	—	0^h 1'			
26.7	—	8	21.42		
16.6	—	15	—	53.15	

Two extreme points missed

25.9	—	36			
18.1	—	43	21.9		
25.5	—	50			

Motion of arm on moving weights from midway to — = 3.13
— to + = 5.72

In these three experiments, the effect of the weight appeared to increase from two to five tenths of a division, on standing an hour ; and the thermometers shewed, that the weights were three or five tenths of a degree warmer than the air close to the case. In the two last experiments, I put a lamp into the room, over night, in hopes of making the air warmer than the weights, but without effect, as the heat of the weights exceeded that of the air more in these two experiments than in the former.

On the evening of October 17, the weights being placed in the midway position, lamps were put under them, in order to warm them ; the door was then shut, and the lamps suffered to burn out. The next morning it was found, on moving the weights to the negative position, that they were 7°.5 warmer than the air near the case. After they had continued an hour in that position, they were found to have cooled 1°.5, so as to be only 6° warmer than the air. They were then moved to the positive position ; and in both positions the arm was drawn aside about four divisions more, after the weights had remained an hour in that position, than it was at first.

May 22, 1798. The experiment was repeated in the same manner, except that the lamps were made so as to burn only a short time, and only two hours were suffered to elapse before the weights were moved. The weights were now found to be scarcely 2° warmer than the case ; and the arm was drawn aside about two divisions more, after the weights had remained an hour in the position they were moved to, than it was at first.

On May 23, the experiment was tried in the same manner, except that the weights were cooled by laying ice on them ; the ice being confined in its place by tin plates, which, on moving the weights, fell to the ground, so as not to be in the way. On moving the weights to the negative position, they were found

79

to be about 8° colder than the air, and their effect on the arm seemed now to diminish on standing, instead of increasing, as it did before ; as the arm was drawn aside about 2½ divisions less, at the end of an hour after the motion of the weights, than it was at first.

It seems sufficiently proved, therefore, that the effect in question is produced, as above explained, by the difference of temperature between the weights and case; for in the 6th, 8th, and 9th* experiments, in which the weights were not much warmer than the case, their effect increased but little on standing ; whereas, it increased much, when they were much warmer than the case, and decreased much, when they were much cooler.

It must be observed, that in this apparatus, the box in which the balls play is pretty deep, and the balls hang near the bottom of it, which makes the effect of the current of air more sensible than it would otherwise be, and is a defect which I intend to rectify in some future experiments.

EXPERIMENT IX. APRIL 29

Weights in positive position

Extreme points	Divisions	Time			Point of rest	Time of middle of vibration		
34.7 35. 34.65	—	—			34.84			

Weights moved to negative position

23.8								
	28 29	11ʰ	18′	29″ ⎱ 58 ⎰	—	11ʰ	18′	43″
33.2	—		—		28.52			
	29 28		25	27 ⎱ 57 ⎰	—		25	40
23.9	—		—		28.25			
32.	—		—		28.01			
24.15	—		—		27.82			
31.	—		—		27.63			
24.4	—		—		27.55			
30.4	—		—		27.47			
	28 27	0	7	4 ⎱ 53 ⎰	—	0	7	26
24.7								

Motion of arm   = 6.32
Time of vibration = 6′ 58″

---

* [*This is evidently a misprint for* 6th, 7th, *and* 8th.]

EXPERIMENT X. MAY 5

Weights in positive position

Extreme points	Divisions	Time			Point of rest	Time of middle of vibration			Difference	
34.5										
33 5	—	—			33.97					
34.4										

Weights moved to negative position

22.3										
	28	10ʰ	43'	42″ ⎬	—	10ʰ	43'	36″		
	29		44	6						
33.2	—		—		27 82		—		7'	0″
	28		50	33 ⎬	—		50	36		
	27		51	0						
22.6	—		—		27.72					
32.5	—		—		27.7					
23.2	—		—		27.58					
31.45	—		—		27.4					
23.5	—		—		27.28					
	27	11	25	20 ⎬	—	11	25	24		
	28			58						
30.7	—		—		27.21		—		7	3
	28		32	0 ⎬	—		32	27		
	27		32	40						
23.95	—		—		27.21		—		6	56
	27		39	19 ⎬	—		39	23		
	28		40	2						
30.25										

Motion of arm = 6.15
Time of vibration = 6' 59″

EXPERIMENT XI. MAY 6

Weights in positive position

Extreme points	Divisions	Time			Point of rest	Time of middle of vibration		
34.9								
34.1	—	—			34.47			
34.8	—	—			34.49			
34 25								

Weights moved to negative position

23.3								
	28	9ʰ	59'	59″ ⎬	—	10ʰ	0'	8″
	29	10	0	27				
33.3	—		—		28.42			
	29		6	52 ⎬	—		7	5
	27		7	51				
23 8	—		—		28 35			

F   81

Extreme points	Divisions	Time			Point of rest	Time of middle of vibration
32.5	—	—			28.3	
24.4						
missed						
24.8						
31.3	—	—			28.17	
	29	10ʰ 48′	37″ }		—	10ʰ 49′ 8″
	28	49	21 }			
25.3	—	—			28.2	
	28	56	8 }		—	56 13
	29		56 }			
30.9						

Motion of arm   = 6.07
Time of vibration = 7′ 1″

In the three foregoing experiments, the index was purposely moved so that, before the beginning of the experiment, the balls rested as near the sides of the case as they could, without danger of touching it; for it must be observed, that when the arm is at 35, they begin to touch. In the two following experiments, the index was in its usual position.

EXPERIMENT XII. MAY 9

Weights in negative position

Extreme points	Divisions	Time			Point of rest	Time of middle of vibration
	17.4	9ʰ 45′	0″			
	17.4	58	0			
	17.4	10 8	0			
	17.4	10	0		17.4	

Weights moved to positive position

Extreme points	Divisions	Time			Point of rest	Time of middle of vibration
28.85						
	24	20	50 }		—	10ʰ 20′ 59″
	22	21	46 }			
18.4	—	—			23.49	
28.3	—	—			23.57	
19.3	—	—			23 67	
27.8	—	—			23.72	
20.	—	—			23.8	
27.4	—	—			23.83	
	24	11 3	13 }		—	11 3 14
	23		54 }			
20.55	—	—			23.87	
	23	9	45 }		—	10 18
	24	10	28 }			
27.						

Motion of arm   = 6.09
Time of vibration = 7′ 3″

EXPERIMENT XIII. MAY 25

Weights in negative position

Extreme points	Divisions	Time			Point of rest	Time of middle of vibration		
16. 18.3 16.2	—	—			17.2			

Weights moved to positive position

Extreme points	Divisions	Time			Point of rest	Time of middle of vibration		
29.6	25 24	10^h 22′	22″ ⎱ 45 ⎰		—	10^h 22′ 56″		
17.4	— 23 24		29 59 ⎱ 30 23 ⎰		23.32 —	30 3		
28.9	— 24 23		36 58 ⎱ 37 24 ⎰		23.4 —	37 7		
18.4	— 23 24		44 3 ⎱ 31 ⎰		23.52 —	44 14		
28.4 19.3 27.8	— — — 24 23	11 5	26 ⎱ 6 1 ⎰		23.62 23.7 23.7 —	11 5 31		
19.9	— 23 24		12 12 ⎱ 50 ⎰		23.72 —	12 35		
27.3								

Weights moved to negative position

Extreme points	Divisions	Time			Point of rest	Time of middle of vibration		
13.5 21.8	— 18 17		37 34 ⎱ 38 10 ⎰		17.75 —	37 39		
13.9	— 17 18		44 26 ⎱ 45 4 ⎰		17.67 —	44 45		
21.1 14.4 20.5 14.7 20.	— — — — — 18 17	0 19	57 ⎱ 20 52 ⎰		17.62 17.6 17.52 17.47 17.42 —	0 20 24		
15.	— 17 18		27 15 ⎱ 28 15 ⎰		17.37 —	27 30		
19.5								

Motion of the arm on moving weights from − to + = 6.12

\ + to − = 5.97

Time of vibration at + \ \ \ \ \ \ \ \ \ \ \ \ = 7′ 6″

\ \ \ \ \ \ \ \ \ \ \ \ \ − \ \ \ \ \ \ \ \ \ \ \ = 7′ 7″

Experiment XIV. May 26

Weights in negative position

Extreme points	Divisions	Time	Point of rest	Time of middle of vibration
	16.1	9ʰ 18 0″		
	16.1	24 0		
	16.1	46 0		
	16.1	49 0	16.1	

Weights moved to positive position

Extreme points	Divisions	Time	Point of rest	Time of middle of vibration
27.7				
	23	10 0 46 ⎰	—	10ʰ 1′ 1″
	22	1 16 ⎱		
17.3	—	—	22.37	
	22	7 58 ⎰	—	8 5
	23	8 27 ⎱		
27.2	—	—	22.5	
	23	15 2 ⎰	—	15 9
	22	32 ⎱		
18.3	—	—	22.65	
26 8	—	—	22.75	
19.1	—	—	22.85	
26.4	—	—	22.97	
	23	43 40 ⎰	—	43 32
	22	44 22 ⎱		
20.	—	—	23.15	
	22	49 53 ⎰	—	50 41
	23	50 37 ⎱		
26.2				

Weights moved to negative position

Extreme points	Divisions	Time	Point of rest	Time of middle of vibration
12.4				
	16	11 7 53 ⎰	—	11 8 25
	17	8 27 ⎱		
21.5	—	—	17.02	
	17	15 30 ⎰	—	15 27
	16	16 3 ⎱		
12.7	—	—	16.9	
20.7	—	—	16.85	
13.3	—	—	16.82	
20.	—	—	16.72	
13.6	—	—	16.67	
	16	50 33 ⎰	—	50 58
	17	51 19 ⎱		
19.5	—	—	16.65	
	17	57 53 ⎰	—	58 6
	16	58 44 ⎱		
14.				

Motion of arm by moving weights from − to + = 6.27

+ to − = 6.13

Time of vibration at + = 7′ 6″

− = 7′ 6″

In the next experiment, the balls, before the motion of the weights, were made to rest as near as possible to the sides of the case, but on the contrary side from what they did in the 9th, 10th, and 11th experiments.

EXPERIMENT XV.   MAY 27

Weights in negative position

Extreme points	Divisions	Time	Point of rest	Time of middle of vibration
3.9				
3.35	—	—	3.61	
3.85	—	—	3.61	
3.4				

Weights moved to positive position

Extreme points	Divisions	Time	Point of rest	Time of middle of vibration
15.4				
	10	10^h 5′ 59″ ⎫	—	10^h 5′ 56″
	9	6 27 ⎭		
4.8	—	—	9 95	
	9	12 43 ⎫	—	13 5
	10	13 11 ⎭		
14.8	—	—	10.07	
	10	20 24 ⎫	—	20 13
	9	56 ⎭		
5.9	—	—	10.23	
14.35	—	—	10.35	
6.8	—	—	10.46	
13.9	—	—	10.52	
	11	48 30 ⎫	—	48 42
	10	49 11 ⎭		
7.5	—	—	10.6	
	10	55 26 ⎫	—	55 48
	11	56 10 ⎭		
13.5				

Motion of the arm = 6.34
Time of vibration = 7′ 7″

The two following experiments were made by Mr. Gilpin, who was so good as to assist me on the occasion.

EXPERIMENT XVI.   MAY 28

Weights in negative position

Extreme points	Divisions	Time	Point of rest	Time of middle of vibration
22.55				
8 4	—	—	15 09	
21	—	—	14.9	
9.2				

Weights moved to positive position

Extreme points	Divisions	Time	Point of rest	Time of middle of vibration
26.6				
	22	10^h 22′ 53″ ⎱	—	10^h 23′ 15″
	21	23 20 ⎰		
15.8	—	—	21.	
	20	30 7 ⎱	—	30 30
	21	36 ⎰		
25.8	—	—	21.05	
	22	37 23 ⎱	—	37 45
	21	55 ⎰		
16.8	—	—	21.11	
	20	44 29 ⎱	—	45 1
	21	45 4 ⎰		
25.05	—	—	21.11	
	22	51 54 ⎱	—	52 20
	21	52 32 ⎰		
17.57	—	—	21.2	
	21	59 31 ⎱	—	59 34
	22	11 0 13 ⎰		
24.6	—	—	21.28	
	22	6 24 ⎱	—	11 6 49
	21	7 9 ⎰		
18.3				

Motion of the arm = 6.1
Time of vibration = 7′ 16″

EXPERIMENT XVII.   MAY 30

Weights in negative position

Extreme points	Divisions	Time	Point of rest	Time of middle of vibration
	17 2	10^h 19′ 0″		
	17.1	25 0		
	17.07	29 0		
	17.15	40 0		
	17.45	49 0		
	17.42	51 0		
	17.42	11 1 0	17.42	

Weights moved to positive position

Extreme points	Divisions	Time	Point of rest	Time of middle of vibration
28.8				
	24	11 11 23 ⎱	—	11^h 11′ 37″
	23	49 ⎰		
18.1	—	—	23.2	
	22	18 13 ⎱	—	18 42
	23	43 ⎰		
27.8	—	—	23.12	
	24	25 19 ⎱	—	25 40
	23	49 ⎰		
18.8	—	—	23.2	
	23	32 41 ⎱	—	32 43
	24	33 13 ⎰		

27.38	—	—	23.31	
	24	11ʰ 39′ 28″ ⎱	—	11ʰ 39′ 44″
	23	40 3 ⎰		
19.7	—	—	23.44	
	23	46 33 ⎱	—	
	24	47 11 ⎰		46 46
27.	—	—	23.52	
	24	53 36 ⎱	—	
	23	54 17 ⎰		53 48
20.4	—	—	23.57	
	23	0 0 34 ⎱	—	
	24	1 18 ⎰		0 0 55
26.5	—	—	23.55	
	24	7 34 ⎱	—	
	23	8 21 ⎰		7 50
20.8	—	—	23.59	
	23	14 30 ⎱	—	
	24	15 24 ⎰		14 58
26 25				

Weights moved to negative position

13.3				
	17	32 19 ⎱	—	
	18	48 ⎰		32 44
22.4	—	—	17.95	
	18	39 46 ⎱	—	
	17	40 19 ⎰		39 44
13.7	—	—	17.85	
	17	46 26 ⎱	—	
	18	47 0 ⎰		46 48
21.6	—	—	17.72	
	18	53 43 ⎱	—	
	17	54 20 ⎰		53 50
14.	—	—	17.6	
	17	1 0 39 ⎱	—	
	18	1 20 ⎰		1 0 55
20.8	—	—	17.47	
	18	7 39 ⎱	—	
	17	8 21 ⎰		7 59
14.3	—	—	17.37	
	17	14 54 ⎱	—	
	18	15 42 ⎰		15 4
20.1	—	—	17.27	
	18	21 32 ⎱	—	
	17	22 22 ⎰		22 5
14.6				

Motion of the arm on moving weights from − to + = 5.78

+ to − = 5.64

Time of vibration at + = 7′ 2″

= 7′ 3″

ON THE METHOD OF COMPUTING THE DENSITY OF THE EARTH FROM THESE EXPERIMENTS

I shall first compute this, on the supposition that the arm and copper rods have no weight, and that the weights exert no sensible attraction, except on the nearest ball; and shall then examine what corrections are necessary, on account of the arm and rods, and some other small causes.

The first thing is, to find the force required to draw the arm aside, which, as was before said, is to be determined by the time of a vibration.

The distance of the centres of the two balls from each other is 73.3 inches, and therefore the distance of each from the centre of motion is 36.65, and the length of a pendulum vibrating seconds, in this climate, is 39.14; therefore, if the stiffness of the wire by which the arm is suspended is such, that the force which must be applied to each ball, in order to draw the arm aside by the angle A, is to the weight of that ball as the arch of A to the radius, the arm will vibrate in the same time as a pendulum whose length is 36.65 inches, that is, in $\sqrt{\dfrac{36.65}{39.14}}$ seconds; and therefore, if the stiffness of the wire is such as to make it vibrate in N seconds, the force which must be applied to each ball, in order to draw it aside by the angle A, is to the weight of the ball as the arch of $A \times \dfrac{1}{N^2} \times \dfrac{36.65}{39.14}$ to the radius. But the ivory scale at the end of the arm is 38.3 inches from the centre of motion, and each division is $\frac{1}{20}$ of an inch, and therefore subtends an angle at the centre, whose arch is $\frac{1}{766}$; and therefore the force which must be applied to each ball, to draw the arm aside by one division, is to the weight of the ball as $\dfrac{1}{766 N^2} \cdot \dfrac{36.65}{39.14}$ to 1, or as $\dfrac{1}{818 N^2}$ to 1.*

* [Or thus: using the ordinary notation for the simple pendulum vibrating through small arcs, if the force on each ball drawing the arm aside through an arc subtending an angle of A° were $mg \times \dfrac{arc}{radius}$, the arm would vibrate like a pendulum of the same length, and have a period of $\sqrt{\dfrac{36.65}{39.14}}$ seconds, because the period of a pendulum varies as the square root of its length. But the force varies as $\dfrac{1}{(period)^2}$; therefore the force required to draw the arm through A° with

88

THE LAWS OF GRAVITATION

The next thing is, to find the proportion which the attraction of the weight on the ball bears to that of the earth thereon, supposing the ball to be placed in the middle of the case, that is, to be not nearer to one side than the other. When the weights are approached to the balls, their centres are 8.85 inches from the middle line of the case; but, through inadvertence, the distance, from each other, of the rods which support these weights, was made equal to the distance of the centres of the balls from each other, whereas it ought to have been somewhat greater. In consequence of this, the centres of the weights are not exactly opposite to those of the balls, when they are approached together; and the effect of the weights, in drawing the arm aside, is less than it would otherwise have been, in the triplicate ratio of $\dfrac{8.85}{36.65}$ to the chord of the angle whose sine is $\dfrac{8.85}{36.65}$, or in the triplicate ratio of the cosine of $\frac{1}{2}$ this angle to the radius, or in the ratio of .9779 to 1.*

period $N'' = mg \times \dfrac{arc \ of \ A°}{36.65} \times \dfrac{36.65}{39 \ 14} \div N^2$. *And the force required to draw the arm through* 1 *scale division with period* N''

$$= mg \times \dfrac{\dfrac{36 \ 65}{38.3} \cdot \dfrac{1}{20}}{36 \ 65} \times \dfrac{36 \ 65}{39 \ 14} \div N^2$$

$$= mg \times \dfrac{1}{766 N^2} \times \dfrac{36 \ 65}{39.14} = mg \times \dfrac{1}{818 N^2} \ ]$$

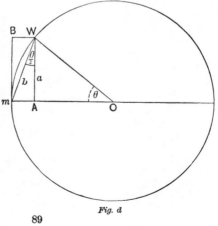

* [*Let* W *be the position of the " weight " of mass* W, B *the position it was intended that it should have, and m that of the "ball" of mass m. The distance* mB, *or* WA, *is* 8.85 *inches, and* OW *and* Om 36 65 *inches. Call* WA *and* Wm *a and b respectively, and* G *the gravitation constant. Then it was intended that the attraction to move the arm should be* $\dfrac{G \cdot W \cdot m}{a^2}$, *but it is* $\dfrac{G \cdot W \cdot m}{b^2} \cdot \dfrac{a}{b}$; *and so is less than was intended in the ratio of* $\dfrac{a^3}{b^3}$ *to* 1, *or of Cos³* $\dfrac{\theta}{2}$ *to* 1.]

Fig. d

Each of the weights weighs 2,439,000 grains, and therefore is equal in weight to 10.64 spherical feet of water ;* and therefore its attraction on a particle placed at the centre of the ball, is to the attraction of a spherical foot of water on an equal particle placed on its surface, as $10.64 \times .9779 \times \left(\dfrac{6}{8.85}\right)^{2}$ to 1. The mean diameter of the earth is 41,800,000 feet ;† and therefore, if the mean density of the earth is to that of water as D to one, the attraction of the leaden weight on the ball will be to that of the earth thereon, as $10.64 \times .9779 \times \left(\dfrac{6}{8.85}\right)^{2}$ to 41,800,000 D : : 1 : 8,739,000 D.‡

* [*That is, is equal to the weight of a sphere of water which can be inscribed in a cube whose volume is 10.64 cu. ft , or we can express the volume of the sphere by the number 10.64, when the unit of volume is that of a sphere of 1 foot in diameter, that is, of $\dfrac{\pi}{6}$ cu ft The radius of a spherical foot of water is, accordingly, 6 inches. Cavendish evidently uses Kirwan's estimate of 253.35 grains to the cu. in. of water.*

The ensuing calculation can be stated thus: Call d and d' the densities of water and of the earth respectively, m the mass of the ball, and G the gravitation constant The volume of the earth in spherical units is $(41\ 800\ 000)^{3}$, *and its radius* $6 \times 41\ 800\ 000$ *inches.*

$$\frac{\textit{Attraction of weight on ball at 8 85 inches}}{\textit{Attraction of earth on ball}} = \frac{\dfrac{G \times 10\ 64 \times d \times m}{(8\ 85)^{2}} \times 9779}{\dfrac{G \times (41\ 800\ 000)^{3} \times d' \times m}{(6 \times 41\ 800\ 000)^{2}}}$$

$$= \frac{.9779 \times 10.64 \times \left(\dfrac{6}{8.85}\right)^{2}}{41\ 800\ 000\ \dfrac{d'}{d}}$$

$$= \frac{1}{8\ 739\ 000\ D} \quad \cdots \quad (1)$$

But we have already found (page 89)

$$\frac{\textit{Force required to draw the arm through 1 div.}}{\textit{Weight of ball}} = \frac{1}{818 N^{2}} \quad \cdots \quad (2)$$

Dividing equation (1) by (2) we have

$$\frac{\textit{Attraction of weight on ball}}{\textit{Force required to draw the arm through 1 div.}} = \frac{818\ N^{2}}{8\ 739\ 000 D} = \frac{N^{2}}{10\ 683 D}$$

$= $ *no. of div. through which the arm is drawn* $\equiv B$ *div.*]

† In strictness, we ought, instead of the mean diameter of the earth, to take the diameter of that sphere whose attraction is equal to the force of gravity in this climate; but the difference is not worth regarding.

‡ [*Hutton has pointed out (54) that this number should be 8,740,000; but it will not make any appreciable change in the value of* D.]

It is shewn, therefore, that the force which must be applied to each ball, in order to draw the arm one division out of its natural position, is $\dfrac{1}{818\,N^2}$ of the weight of the ball; and, if the mean density of the earth is to that of water as D to 1, the attraction of the weight on the ball is $\dfrac{1}{8,739,000\,D}$ of the weight of that ball; and therefore the attraction will be able to draw the arm out of its natural position by $\dfrac{818\,N^2}{8,739,000\,D}$ or $\dfrac{N^2}{10,683\,D*}$ divisions; and therefore, if on moving the weights from the midway to a near position the arm is found to move B divisions, or if it moves 2 B divisions on moving the weights from one near position to the other, it follows that the density of the earth, or D, is $\dfrac{N^2}{10,683\,B}$.

We must now consider the corrections† which must be applied to this result; first, for the effect which the resistance of the arm to motion has on the time of the vibration : 2d, for the attraction of the weights on the arm : 3d, for their attraction on the farther ball : 4th, for the attraction of the copper rods on the balls and arm : 5th, for the attraction of the case on the balls and arm : and 6th, for the alteration of the attraction of the weights on the balls, according to the position of the arm, and the effect which that has on the time of vibration. None of these corrections, indeed, except the last, are of much signification, but they ought not entirely to be neglected.

As to the first, it must be considered, that during the vibrations of the arm and balls, part of the force is spent in accelerating the arm ; and therefore, in order to find the force required to draw them out of their natural position, we must find the proportion which the forces spent in accelerating the arm and balls bear to each other.

Let EDC*edc* (Fig. 4) be the arm. B and *b* the balls. C*s* the suspending wire. The arm consists of 4 parts ; first, a deal rod D*cd*, 73.3 inches long ; 2d, the silver wire DC*d*, weighing 170 grains ; 3d, the end pieces DE and *ed*, to which the ivory

---

* [*This number should be* 10,685. *See last note.*]

† [*For a discussion of these corrections, similar to that of Cavendish, but with modern mathematical treatment, see Reich* (67)]

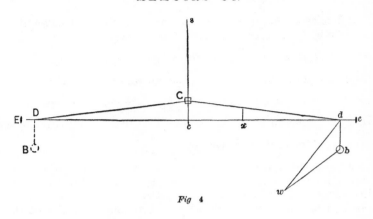

Fig 4

vernier is fastened, each of which weighs 45 grains ; and 4th, some brass work Cc, at the centre. The deal rod, when dry, weighs 2320 grains, but when very damp, as it commonly was during the experiments, weighs 2400 ; the transverse section is

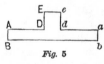

Fig. 5

of the shape represented in Fig. 5 ; the thickness BA, and the dimensions of the part DEed, being the same in all parts ; but the breadth Bb diminishes gradually, from the middle to the ends. The area of this section is .33 of a square inch at the middle, and .146 at the end ; and therefore, if any point x (Fig. 4) is taken in cd, and $\dfrac{cx}{cd}$ is called x, this rod weighs $\dfrac{2400 \times .33}{73.3 \times .238}$ per inch at the middle ; $\dfrac{2400 \times .146}{73.3 \times .238}$ at the end, and $\dfrac{2400}{73.3} \times \dfrac{.33 - .184\,x}{.238} = \dfrac{3320 - 1848\,x}{73.3}$ at x ; and therefore, as the weight of the wire is $\dfrac{170}{73.3}$ per inch, the deal rod and wire together may be considered as a rod whose weight at $x = \dfrac{3490 - 1848\,x}{73.3}$ per inch.

But the force required to accelerate any quantity of matter placed at x, is proportional to x^2 ; that is, it is to the force required to accelerate the same quantity of matter placed at d as x^2 to 1 ; and therefore, if cd is called l, and x is supposed to flow, the fluxion of the force required to accelerate the deal rod and wire is proportional to $\dfrac{x^2 l\dot{x} \times (3490 - 1848\,x)}{73.3}$, the fluent

of which, generated while x flows from c to d,

$$= \frac{l}{73.3} \times \left(\frac{3490}{3} - \frac{1848}{4} \right) = 350 \; ;$$

so that the force required to accelerate each half of the deal rod and wire, is the same as is required to accelerate 350 grains placed at d.

The resistance to motion of each of the pieces de, is equal to that of 48 grains placed at d; as the distance of their centres of gravity from C is 38 inches. The resistance of the brass work at the centre may be disregarded; and therefore the whole force required to accelerate the arm, is the same as that required to accelerate 398 grains placed at each of the points D and d.

Each of the balls weighs 11,262 grains, and they are placed at the same distance from the centre as D and d; and therefore, the force required to accelerate the balls and arm together, is the same as if each ball weighed 11,660, and the arm had no weight; and therefore, supposing the time of a vibration to be given, the force required to draw the arm aside, is greater than if the arm had no weight, in the proportion of 11,660 to 11,262, or of 1.0353 to 1.

To find the attraction of the weights on the arm, through d draw the vertical plane dwb perpendicular to Dd, and let w be the centre of the weight, which, though not accurately in this plane, may, without sensible error, be considered as placed therein, and let b be the centre of the ball; then wb is horizontal and $= 8.85$, and db is vertical and $= 5.5$; let $wd = a$, $wb = b$, and let $\dfrac{dx}{dc}$, or $1 - x$, $= z$; then the attraction of the weight on a particle of matter at x, in the direction bw, is to its attraction on the same particle placed at $b :: b^3 : (a^2 + z^2 l^2)^{\frac{3}{2}}$, or is proportional to $\dfrac{b^3}{(a^2 + z^2 l^2)^{\frac{3}{2}}}$, and the force of that attraction to move the arm, is proportional to $\dfrac{b^3 \times (1 - z)}{(a^2 + z^2 l^2)^{\frac{3}{2}}}$, and the weight of the deal rod and wire at the point x, was before said to be $\dfrac{3490 - 1848\,x}{73.3} = \dfrac{1642 + 1848\,z}{73.3}$ per inch; and therefore, if dx flows, the fluxion of the power to move the arm

$$= lz \times \frac{1642 + 1848\,z}{73.3} \times \frac{b^3 \times (1-z)}{(a^2 + z^2 l^2)^{\frac{3}{2}}} = \dot{z} \times (821 + 924\,z) \times \frac{b^3 \times (1-z)}{(a^2 + l^2 z^2)^{\frac{3}{2}}}$$

$$= \frac{b^3 z \times (821 + 103\,z - 924\,z^2)}{(a^2 + l^2 z^2)^{\frac{3}{2}}} = \frac{b^3 \dot{z} \times \left(821 + 103\,z + \dfrac{924\,a^2}{l^4}\right)}{(a^2 + l^2 z^2)^{\frac{3}{2}}} -$$

$$- \frac{924 b^3 z \times \left(\dfrac{a^2}{l^4} + z^2\right)}{(a^2 + l^2 z^2)^{\frac{3}{2}}} \; ; \; \text{which, as } \frac{a^2}{l^2} = .08,$$

$$= \frac{b^3 z \times (895 + 103\,z)}{(a^2 + l^2 z^2)^{\frac{3}{2}}} - \frac{924\ b^3 z}{l^2 \sqrt{a^2 + l^2 z^2}}. \quad \text{The fluent of this}$$

$$= \frac{895\ b^3 z}{a^2 \sqrt{a^2 + l^2 z^2}} - \frac{103\ b^3}{l^2 \sqrt{a^2 + l^2 z^2}} + \frac{103\ b^3}{l^2 a} - \frac{924\ b^3}{l^3} \log \frac{lz + \sqrt{a^2 + l^2 z^2}}{a},$$

and the force with which the attraction of the weight, on the nearest half of the deal rod and wire, tends to move the arm, is proportional to this fluent generated while z flows from 0 to 1, that is, to 128 grains.

The force with which the attraction of the weight on the end piece de tends to move the arm, is proportional to $47 \times \dfrac{b^3}{a^2}$ [*approximately*], or 29 grains; and therefore the whole power of the weight to move the arm, by means of its attraction on the nearest part thereof, is equal to its attraction on 157 grains placed at b, which is $\dfrac{157}{11262}$, or .0139 of its attraction on the ball.*

It must be observed, that the effect of the attraction of the weight on the whole arm is rather less than this, as its attraction on the farther half draws it the contrary way; but, as the attraction on this is small, in comparison of its attraction on the nearer half, it may be disregarded.

The attraction of the weight on the furthest ball, in the direction bw, is to its attraction on the nearest ball :: $wb^3 : wB^3$†
:: .0017 : 1 ; and therefore the effect of the attraction of the weight on both balls, is to that of its attraction on the nearest ball :: .9983 : 1.

---

* [*A few minor misprints in the last two paragraphs in the original paper have been corrected. A recalculation seems to give 142.5 instead of 128, and 28 instead of 29, grains; this would change the value of* D *by 1 part in* 1000.]

† [*This is erroneously printed in the original as* $wd^3 : wD^3$.]

To find the attraction of the copper rod on the nearest ball, let b and w (Fig. 6) be the centres of the ball and weight, and ea the perpendicular part of the copper rod, which consists of two parts, ad and de. ad weighs 22,000 grains, and is 16 inches long, and is nearly bisected by w. de weighs 41,000, and is 46 inches long. wb is 8.85 inches, and is perpendicular to ew.

Now, the attraction of a line ew, of uniform thickness, on b, in the direction bw, is to that of the same quantity of matter placed at $w :: bw : eb;$ and therefore the attraction of the part da equals that of $\dfrac{22,000 \times wb}{db}$, or 16,300, placed at $w;$ and the attraction of de equals that of $41,000 \times \dfrac{ew}{ed} \times \dfrac{bw}{be} - 41,000 \times \dfrac{dw}{ed} \times \dfrac{bw}{bd}$, or 2500, placed at the same point ; so that the attraction of the perpendicular part of the copper rod on b, is to that of the weight thereon, as 18,800 : 2,439,-000, or as .00771 to 1. As for the attraction of the inclined part of the rod and wooden bar, marked

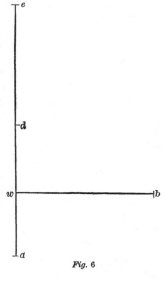

Fig. 6

Pr and rr in Fig. 1, it may safely be neglected, and so may the attraction of the whole rod on the arm and farthest ball ; and therefore the attraction of the weight and copper rod, on the arm and both balls together, exceeds the attraction of the weight on the nearest ball, in the proportion of .9983 + .0139 + .0077 to one, or of 1.0199 to 1.

The next thing to be considered, is the attraction of the mahogany case. Now it is evident, that when the arm stands at the middle division, the attractions of the opposite sides of the case balance each other, and have no power to draw the arm either way. When the arm is removed from this division, it is attracted a little towards the nearest side, so that the force required to draw the arm aside is rather less than it would otherwise be ; but yet, if this force is proportional to the distance of the arm from the middle division, it makes no error in the result; for, though the attraction will draw the arm aside more than it would otherwise do, yet, as the accelerating force by which the arm is made to vibrate is diminished in the same proportion, the square of the time of a vibration will be increased in the same proportion as the space by which the arm is drawn aside, and therefore the result will be the same as if the case exerted no attraction; but, if the attraction of the case is not proportional to the distance of the arm from the middle point, the ratio in which the accelerating force is diminished is different in different parts of the vibration, and the square of the time of a vibration will not be increased in the same proportion as the quantity by which the arm is drawn aside, and therefore the result will be altered thereby.

On computation, I find that the force by which the attraction draws the arm from the centre is far from being proportional to the distance, but the whole force is so small as not to be worth regarding; for, in no position of the arm does the attraction of the case on the balls exceed that of $\frac{1}{6}$th of a spheric inch of water, placed at the distance of one inch from the centre of the balls; and the attraction of the leaden weight equals that of 10.6 spheric feet of water placed at 8.85 inches, or of 234 spheric inches placed at 1 inch distance; so that the attraction of the case on the balls can in no position of the arm exceed $\frac{1}{1170}$ of that of the weight. The computation is given in the Appendix.

It has been shown, therefore, that the force required to draw the arm aside one division, is greater than it would be if the arm had no weight, in the ratio of 1.0353 to 1, and therefore $=\dfrac{1.0353}{818\,N^2}$ of the weight of the ball; and moreover, the attraction of the weight and copper rod on the arm and both balls together, exceeds the attraction of the weight on the nearest ball, in the ratio of 1.0199 to 1, and therefore $=\dfrac{1.0199}{8,739,000D}$ of the weight of the ball; consequently D is really equal to $\dfrac{818N^2}{1.0353}$ $\times\dfrac{1.0199}{8,739,000B}$, or $\dfrac{N^2}{10,844B}*$, instead of $\dfrac{N^2}{10,683B}$, as by the former computation. It remains to be considered how much this is affected by the position of the arm.

* [*This should be* 10,846; *see note on pp.* 90 *and* 91.]

Suppose the weights to be approached to the balls; let W (Fig. 7) be the centre of one of the weights; let M be the centre of the nearest ball at its mean position, as when the arm is at 20 divisions; let B be the point which it actually rests at; and let A be the point which it would rest at, if the weight was removed; consequently, AB is the space by which it is drawn aside by means of the attraction; and let Mβ be the space by which it would be drawn aside, if the attraction on it was

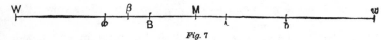

Fig. 7

the same as when it is at M. But the attraction at B is greater than at M, in the proportion of $WM^2 : WB^2$; and therefore, $AB = M\beta \times \dfrac{WM^2}{WB^2} = M\beta \times \left(1 + \dfrac{2MB}{MW}\right)$, very nearly.

Let now the weights be moved to the contrary near position, and let w be now the centre of the nearest weight, and b the point of rest of the centre of the ball; then $Ab = M\beta \times \left(1 + \dfrac{2Mb}{MW}\right)$, and $Bb = M\beta \times \left(2 + \dfrac{2Mb}{MW} + \dfrac{2MB}{MW}\right) = 2M\beta \times \left(1 + \dfrac{Bb}{MW}\right)$; so that the whole motion Bb is greater than it would be if the attraction on the ball was the same in all places as it is at M, in the ratio of $1 + \dfrac{Bb}{MW}$ to one; and, therefore, does not depend sensibly on the place of the arm, in either position of the weights, but only on the quantity of its motion, by moving them.

This variation in the attraction of the weight, affects also the time of vibration; for, suppose the weights to be approached to the balls, let W be the centre of the nearest weight; let B and A represent the same things as before; and let x be the centre of the ball, at any point of its vibration; let AB represent the force with which the ball, when placed at B, is drawn towards A by the stiffness of the wire; then, as B is the point of rest, the attraction of the weight thereon will also equal AB; and, when the ball is at x, the force with which it is drawn towards A, by the stiffness of the wire, $= Ax$, and that with which it is drawn in the contrary direction, by the attraction, $= AB \times \dfrac{WB^2}{Wx^2}$; so that the actual force by which it is drawn

towards $A = Ax - \dfrac{AB \times WB^2}{Wx^2} = AB + Bx - AB \times \left(1 + \dfrac{2Bx}{WB}\right) = Bx$
$- \dfrac{2Bx \times AB}{WB}$, very nearly. So that the actual force with which the ball is drawn towards the middle point of the vibration, is less than it would be if the weights were removed, in the ratio of $1 - \dfrac{2AB}{WB}$ to one, and the square of the time of a vibration is increased in the ratio of 1 to $1 - \dfrac{2AB}{WB}$; which differs very little from that of $1 + \dfrac{Bb}{MW}$ to 1, which is the ratio in which the motion of the arm, by moving the weights from one near position to the other, is increased.

The motion of the ball answering to one division of the arm $= \dfrac{36.65*}{20 \times 38.3}$; and, if MB† is the motion of the ball answering to d divisions on the arm, $\dfrac{MB}{WM} = \dfrac{36.65d}{20 \times 38.3 \times 8.85} = \dfrac{d}{185}$; and therefore, the time of vibration, and motion of the arm, must be corrected as follows :

If the time of vibration is determined by an experiment in which the weights are in the near position, and the motion of the arm, by moving the weights from the near to the midway position, is d divisions, the observed time must be diminished in the subduplicate ratio of $1 - \dfrac{2d}{185}$ to 1, that is, in the ratio of $1 - \dfrac{d}{185}$ to 1; but, when it is determined by an experiment in which the weights are in the midway position, no correction must be applied.

To correct the motion of the arm caused by moving the weights from a near to the midway position, or the reverse, observe how much the position of the arm differs from 20 divisions, when the weights are in the near position ; let this be n divisions, then, if the arm at that time is on the same side of the division of 20 as the weight, the observed motion must be

---

* [*This number, 36.65, here, and again in the next line, is erroneously printed in Cavendish's memoir as 36.35.*]

† [*In the original this is erroneously printed as m*B.]

diminished by the $\frac{2n}{185}$ part of the whole; but, otherwise, it must be as much increased.

If the weights are moved from one near position to the other, and the motion of the arm is $2d$ divisions, the observed motion must be diminished by the $\frac{2d}{185}$ part of the whole.

If the weights are moved from one near position to the other, and the time of vibration is determined while the weights are in one of those positions, there is no need of correcting either the motion of the arm, or the time of vibration.*

Conclusion

The following table contains the result of the experiments

Exper	Mot weight	Mot arm	Do corr	Time vib	Do corr	Density
1 {	m to +	14.32	13.42		—	5.5
	+ to m.	14.1	13.17	14′ 55″	—	5.61
2 {	m. to +	15.87	14.69	—	—	4.88
	+ to m.	15.45	14.14	14 42	—	5.07
3 {	+ to m.	15.22	13.56	14 39	—	5.26
	m. to +	14.5	13.28	14 54	—	5.55
4 {	m. to +	3.1	2.95		6′ 54″	5.36
	+ to −	6.18	—	7 1	—	5.29
	− to +	5.92	—	7 3	—	5.58
5 {	+ to −	5.9	—	7 5	—	5.65
	− to +	5.98	—	7 5	—	5.57
6 {	m. to −	3.03	2.9			5.53
	− to +	5.9	5.71	7 4		5.62
7 {	m. to −	3.15	3.03	by	6 57	5.29
	− to +	6.1	5.9	mean		5.44
8 {	m. to −	3.13	3.00			5.34
	− to +	5.72	5.54			5.79
9	+ to −	6.32	—	6 58	—	5.1
10	+ to −	6.15	—	6 59	—	5.27
11	+ to −	6.07	—	7 1	—	5.39
12	− to +	6.09	—	7 3	—	5.42
13 {	− to +	6.12	—	7 6	—	5.47
	+ to −	5.97	—	7 7	—	5.63
14 {	− to +	6.27	—	7 6	—	5.34
	+ to −	6.13	—	7 6	—	5.46
15	− to +	6.34	—	7 7	—	5.3
16	− to +	6.1	—	7 16	—	5.75
17 {	− to +	5.78	—	7 2	—	5.68
	+ to −	5.64	—	7 3	—	5.85

* [*The corrections neutralize each other, since they are the same for* N^2 *and* B, *whose ratio enters into the expression for* D.]

From this table it appears, that though the experiments agree pretty well together, yet the difference between them, both in the quantity of motion of the arm and in the time of vibration, is greater than can proceed merely from the error of observation. As to the difference in the motion of the arm, it may very well be accounted for, from the current of air produced by the difference of temperature ; but, whether this can account for the difference in the time of vibration, is doubtful. If the current of air was regular, and of the same swiftness in all parts of the vibration of the ball, I think it could not; but, as there will most likely be much irregularity in the current, it may very likely be sufficient to account for the difference.

By a mean of the experiments made with the wire first used, the density of the earth comes out 5.48 * times greater than that of water ; and by a mean of those made with the latter wire, it comes out the same ; and the extreme difference of the results of the 23 observations made with this wire, is only .75 ; so that the extreme results do not differ from the mean by more than .38, or $\frac{1}{14}$ of the whole, and therefore the density should seem to be determined hereby, to great exactness. It, indeed, may be objected, that as the result appears to be influenced by the current of air, or some other cause, the laws of which we are not well acquainted with, this cause may perhaps act always, or commonly, in the same direction, and thereby make a considerable error in the result. But yet, as the experiments were tried in various weathers, and with considerable variety in the difference of temperature of the weights and air, and with the arm resting at different distances from the sides of the case, it seems very unlikely that this cause should act so uniformly in the same way, as to make the error of the mean result nearly equal to the difference between this and the ex-

---

* [*This should be* 5 31   *Had the third number in the column of densities been* 5.88, *instead of* 4.88, *the average would have been as Cavendish gave it. But Baily* (79, *p.* 90) *recalculated the densities from Cavendish's data, and found* 4.88 *to be correct. Curiously enough Cavendish made the same error in deducing the mean result of the whole number of experiments. It should be* 5.448, *not* 5 48 (*which would be had by putting* 5.88 *in place of* 4.88), *with a probable error of* .033. *The mean result of the last* 23 *observations is* 5.48. *The greatest difference of the single results from one another is* .97 ; *and the extreme result differs from the mean of all by* .57, *or* $\frac{1}{9.8}$ *of the whole. For an account of Hutton's reflections on Cavendish's "pretty and amusing little experiment," see his paper* (45 *and* 54) ; *they are referred to in this volume on page* 105.]

treme ; and, therefore, it seems very unlikely that the density of the earth should differ from 5.48 by so much as $\frac{1}{14}$ of the whole.*

Another objection, perhaps, may be made to these experiments, namely, that it is uncertain whether, in these small distances, the force of gravity follows exactly the same law as in greater distances. There is no reason, however, to think that any irregularity of this kind takes place, until the bodies come within the action of what is called the attraction of cohesion, and which seems to extend only to very minute distances. With a view to see whether the result could be affected by this attraction, I made the 9th, 10th, 11th, and 15th experiments, in which the balls were made to rest as close to the sides of the case as they could ; but there is no difference to be depended on, between the results under that circumstance, and when the balls are placed in any other part of the case.

According to the experiments made by Dr. Maskelyne, on the attraction of the hill Schehallien, the density of the earth is $4\frac{1}{2}$ times that of water ; which differs rather more from the preceding determination than I should have expected. But I forbear entering into any consideration of which determination is most to be depended on, till I have examined more carefully how much the preceding determination is affected by irregularities whose quantity I cannot measure.

* [*See note on page* 100.]
101

APPENDIX

ON THE ATTRACTION OF THE MAHOGANY CASE ON THE BALLS

THE first thing is, to find the attraction of the rectangular plane $ck\beta b$ (Fig 8) on the point a, placed in the line ac perpendicular to this plane.

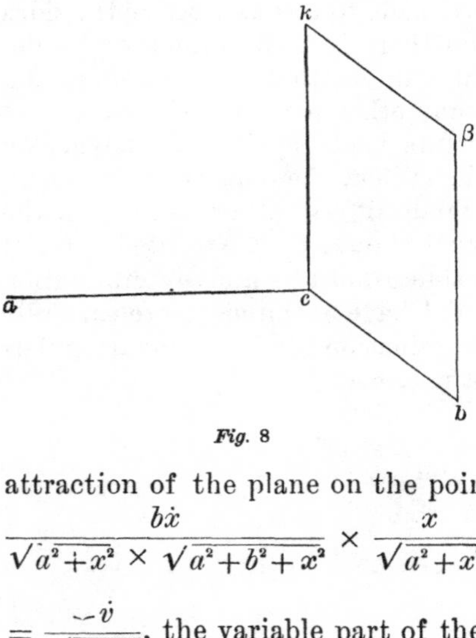

Let $ac = a$, $ck = b$, $cb = x$, and let $\dfrac{a^2}{a^2 + x^2} = w^2$, and $\dfrac{b^2}{a^2 + x^2} = v^2$, then the attraction of the line $b\beta$ on a, in the direction ab, $= \dfrac{b\beta}{ab \times a\beta}$; and therefore, if cb flows, the fluxion of the attraction of the plane on the point a, in the direction cb, $=$

$$\frac{b\dot{x}}{\sqrt{a^2+x^2} \times \sqrt{a^2+b^2+x^2}} \times \frac{x}{\sqrt{a^2+x^2}} = \frac{-b\dot{w}}{w\sqrt{b^2 + \dfrac{a^2}{w^2}}} = \frac{-b\dot{w}}{\sqrt{b^2w^2 + a^2}}$$

$$= \frac{-\dot{v}}{\sqrt{1+v^2}},$$ the variable part of the fluent of which $= -\log (v + \sqrt{1+v})^2$, and therefore the whole attraction $= \log \left(\dfrac{ck+ak}{ac} \times \dfrac{ab}{b\beta+a\beta} \right)$; so that the attraction of the plane, in the direction cb, is found readily by logarithms, but I know no way of finding its attraction in the direction ac, except by an infinite series.[*]

Fig. 8

---

[*] [*Playfair has given an expression in finite terms for this attraction on pp. 225-8 of his paper in the* Trans Roy Soc Edin , vol. 6, 1812. *pp* 187-243,

The two most convenient series I know, are the following :

First Series. Let $\frac{b}{a} = \pi$, and let A=arc whose tang. is π, B $=A-\pi$, $C=B+\frac{\pi^3}{3}$, $D=C-\frac{\pi^5}{5}$, etc.   Then the attraction in the direction $ac = \sqrt{1-w^2} \times \left(A + \frac{Bw^2}{2} + \frac{3Cw^4}{2\cdot4} + \frac{3\cdot5Dw^6}{2\cdot4\cdot6}, \text{etc.} \right).$*

For the second series, let A=arc whose tang. $=\frac{1}{\pi}$, $B=A-\frac{1}{\pi}$, $C=B+\frac{1}{3\pi^3}$, $D=C-\frac{1}{5\pi^5}$, etc.   Then the attraction = arc.$90°$ $-\sqrt{(1+v^2)\left(A - \frac{Bv^2}{2} + \frac{3Cv^4}{2\cdot4} - \frac{3\cdot5Dv^6}{2\cdot4\cdot6}, \text{etc.} \right)}.$

It must be observed, that the first series fails when π is greater than unity, and the second, when it is less ; but, if b is taken equal to the least of the two lines ck and cb, there is no case in which one or the other of them may not be used conveniently.

By the help of these series, I computed the following table :

	.1962	.3714	.5145	6248	7071	.7808	.8575	.9285	9815	1.
.1962	.00001									
.3714	.00039	.00148								
.5145	.00074	.00277	.00521							
.6248	00110	.00406	00778	01183						
.7071	00140	.00522	01008	.01525	02002					
.7808	.00171	.00637	.01245	.01896	.02405	03247				
.8575	00207	00772	01522	02339	.03116	03964	.05057			
.9285	00244	.00910	.01810	.02807	03778	.04867	06319	.08119		
.9815	.00271	01019	.02084	03193	.04368	05639	07478	09931	.12849	
1.	.00284	.01054	02135	03347	.04560	.05975	07978	10789	.14632	.19612

Find in this table, with the argument $\frac{ck}{ak}$ at top, and the argument $\frac{cb}{ab}$ in the left-hand column, the corresponding logarithm ;

entitled " Of the Solids of Greatest Attraction, or those which, among all the Solids that have certain Properties, attract with the greatest Force in a given Direction."]

  * [*In the last term of the series the coefficient* D *was omitted in the original*]

then add together this logarithm, the logarithm of $\frac{ck}{ak}$, and the logarithm of $\frac{cb}{ab}$; the sum is the logarithm of the attraction.

To compute from hence the attraction of the case on the

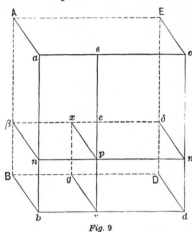

Fig. 9

ball, let the box DCBA (Fig. 1), in which the ball plays, be divided into two parts, by a vertical section, perpendicular to the length of the case, and passing through the centre of the ball; and, in Fig. 9, let the parallelopiped ABDE*abde* be one of these parts, ABDE being the above-mentioned vertical section; let x be the centre of the ball, and draw the parallelogram $\beta npm\delta x$ parallel to B*bd*D, and *xgrp* parallel to βB*bn*, and bisect $\beta\delta$ in *c*. Now, the dimensions of the box, on the inside, are B*b*=1.75; BD=3.6; Bβ=1.75; and βA=5; whence I find that, if *xc* and βx are taken as in the two upper lines of the following table, the attractions of the different parts are as set down below.

		.75	.5	.25
	$x c$	.75	.5	.25
	βx	1.05	1.3	1.55
Excess of attraction of D*drg* above B*brg*		.2374	.1614	.0813
mdrp above *nbrp*		.2374	.1614	.0813
mesp above *nasp*		.3705	.2516	.1271
Sum of these		.8453	.5744	.2897
Excess of attraction of B*bn*β above D*dm*δ		.5007	.3271	.1606
A*an*β above E*em*δ		.4677	.3079	.1525
Whole attraction of the inside surface of the half box.		.1231	.0606	.0234

It appears, therefore, that the attraction of the box on x increases faster than in proportion to the distance *xc*.

The specific gravity of the wood used in this case is .61, and its thickness is $\frac{3}{4}$ of an inch; and therefore, if the attraction of the outside surface of the box was the same as that of the inside, the whole attraction of the box on the ball, when cx $=.75$, would be equal to $2 \times .1231 \times .61 \times \frac{3}{4}$ cubic inches, or .201 spheric inches of water, placed at the distance of one inch from the centre of the ball. In reality it can never be so great as this, as the attraction of the outside surface is rather less than that of the inside; and, moreover, the distance of x from c can never be quite so great as .75 of an inch, as the greatest motion of the arm is only $1\frac{1}{2}$ inch.

Much has been written concerning the Cavendish experiment; the following references may be consulted to advantage.

Gilbert (40), in 1799, translated the greater part of Cavendish's paper into German for his *Annalen*, adding many explanatory notes. A few years later Brandes (42) gave a fresh mathematical analysis of the experiment, including the equations for the time of swing of the torsion pendulum in the experiment proposed by Muncke (see below). In 1815, the original paper of Cavendish was translated entire into French by M. Chompré (50).

In 1821, Hutton (54) recalculated the results of the experiment after Cavendish's own formulæ, and found, as he thought, a "copious list of errata, some of which are large or important." The mean of the first 6 experiments so corrected is 5.19, and of the other 23 is 5.43; the mean of these two means is 5.31, which Hutton takes as the correct result given by the Cavendish experiment. Baily states, however (79, pp. 92-96), that Hutton himself had fallen into error, and that the computations of Cavendish are correct except in the one detail referred to on page 100 of this volume. Baily gives a very careful criticism of the experiment on pp. 88-91 of his memoir. He remarks that "Cavendish's object, in drawing up his memoir, appears to have been more for the purpose of exhibiting a *specimen* of what he considered to be an excellent method of determining this important inquiry, than of deducing a result that should lay claim to the full confidence of the scientific world." Baily points out that the time was not determined with due accuracy; that the experiments were not arranged in groups, in order to eliminate the error arising from the march of the resting point; and that the distance between the weight and the ball was assumed constant. We shall see later from the accounts of the investigations of Reich, Baily, Cornu and Baille, and Boys how the errors in Cavendish's experiment have been avoided.

Muncke (61, vol. 3, pp. 940–70) has given an account of the experiment and an admirable criticism of it, and compares the result with that obtained by Maskelyne and Hutton. He proposed another method of using the torsion balance to find the mean density of the earth ; he would find the time of vibration with the masses first in the line of the balls and then in a line at right angles to that direction. There would be no deflection to be measured. We have seen above that Brandes gave the theory of this experiment. Investigations of this nature have been made by Reich (83), Eötvös (192) and Braun (193); for accounts of which see the latter part of this volume.

A useful résumé of Cavendish's paper was given by Schmidt (64, vol. 2, pp. 481–7). He formed anew equations from which to derive the value of the density of the earth, and found 5.52 using Cavendish's data.

Another mathematical investigation of the dynamical problem underlying the Cavendish experiment was made by Menabrea (71, 72 and 73), in 1840. It is a very elaborate analysis of the whole problem. He examines the effect of the resistance of the air on the time of vibration, and also shews how to find the mass of the earth supposing that it is composed of spheroidal layers of variable density. In Baily's memoir (79) is another elaborate analysis, by Airy, of the mathematical theory of the investigation. It treats especially of Baily's modification of the Cavendish experiment (reproduced in Routh's *Rigid Dynamics* 1882, pt. 1, pp. 359–364).

An elementary treatment of the problem involved is given by Gosselin (127), and from the formula he arrives at he derives the value of the mean density of the earth as given by Cavendish's experiment, and gets 5.69. A similarly elementary treatment by Babinet (132) gives 5.5.

An excellent account of Cavendish's work is given by Zanotti-Bianco (148½) and by Poynting (185, pp. 40–8) ; in the latter is to be found a diagram showing the closeness of Cavendish's separate results to the mean.

THE LAWS OF GRAVITATION

HENRY CAVENDISH, son of Lord Charles Cavendish and a
nephew of the third Duke of Devonshire. was born at Nice in
1731 and died at London in 1810.   He studied at Cambridge,
and becoming possessed, by the death of an uncle, of a large
fortune he devoted his life unostentatiously to private scientific
studies.   Besides the investigation on gravitational attraction
here reprinted, he is remarkable for his researches in the field
of chemistry, and has been called the " Newton " of that subject.   He worked on the constituents of the atmosphere and
on hydrogen ; he made the first synthesis of water, by burning
hydrogen in air, and found the density of hydrogen to be $\frac{1}{11}$
(instead of $\frac{1}{14}$) of that of air.   He determined the ratio of de-
phlogisticated to phlogisticated air to be about as 1 : 4.   Cavendish also made many researches of great importance in the subject of electricity ; these have been collected and edited by
Clerk-Maxwell.   Perhaps the most important of his electrical
investigations is that which proved that electrostatic attraction
takes place according to the law of the inverse square of the
distance.   He is also the author of several papers on astronomical questions.   Most of his writings are to be found in the
Philosophical Transactions of the period.

HISTORICAL ACCOUNT OF THE EXPERI-
MENTS MADE SINCE THE TIME
OF CAVENDISH

HISTORICAL ACCOUNT OF THE EXPERIMENTS MADE SINCE THE TIME OF CAVENDISH

CARLINI. In 1821, Carlini, director of the Brera observatory at Milan, made a series of experiments at the Hospice on Mt. Cenis in the Alps, to determine the length of the seconds-pendulum (55). He was led to do so from considering that the Alps offered a favourable situation for a determination of the mean density of the earth, and that no pendulum experiments had been made there since the publication of the fictitious ones of Coultaud and Mercier (see p. 47). Carlini compared the time of vibration of a simple pendulum, made after the general style of Borda's, with that of a standard clock whose rate was noted daily. The height of the observing station was 1943 metres above sea-level, in latitude 45° 14′ 10″. The corrected length of the seconds-pendulum reduced to sea-level was found to be 993.708 mm. Matthieu and Biot had found the length of the "decimal" seconds-pendulum at Bordeaux, in lat. 44° 50′ 25″, to be 741.6151 mm. The calculated length at Mt. Cenis would be 741.6421 mm.; or, for the "sexagesimal" seconds-pendulum (100 000 decimal seconds = 86 400 sexagesimal seconds) 993.498 mm. The difference between this length and the observed length is .210 mm., which represents the attraction of the mountain on the pendulum.

The mountain is composed of schist, marble and gypsum, of specific gravities 2.81, 2.86 and 2.32 respectively. Carlini took the average of all three, 2.66, as the mean density of the hill. Assuming that the hill was a segment of a sphere 1 geographical mile in height and had a base of 11 miles in diameter, the attraction was calculated to be 5.020δ, where δ is the specific gravity of the hill. With the same units the attraction of the

111

earth is 14 394Δ, where Δ is the mean density of the earth.*
Whence $\frac{5.020\delta}{14\ 394\Delta}=\frac{.210}{993.498}$, and Δ=4.39.

We cannot place very great confidence in this result, not only
on account of the fact that the extreme value of the length of
the seconds-pendulum varies from the mean by .032 mm. and
only 13 determinations were made; but especially because the
size and density to be assigned to the mountain are largely a
matter of conjecture.

A résumé of Carlini's paper was given by Saigey (56), and
by Schell (135). Excellent accounts of the experiment, with
criticisms of it, have been given by Zanotti-Bianco (148½, pt. 2,
pp. 136–45), Poynting (185, pp. 22–4) and Fresdorf (186½, pp.
8–11).

Sabine (58 and 82, notes, p. 47) remarks that Biot and Carlini had not properly reduced to vacuo the observed pendulum
lengths, and states that the corrected length of the seconds-
pendulum on Mt. Cenis is 39.0992 in. From the observations
made in the Formentera-Dunkirk survey he finds by interpolation a pendulum length of 39.1154 in for the latitude of the
Hospice. The difference between the observed and calculated
lengths is .0162 in. The difference calculated from the inverse-
square law is .0238 in. With Carlini's data and equations he
derives 4.77 for the value of Δ.

Schmidt (64, vol. 2, p. 480) gives a concise account of the
theory of the experiment, and remarks that Carlini made an
error in determining the attraction of a spherical segment.
Making the necessary correction he finds Δ = 4.837, a result
not far from that of the Schehallien experiment.

In 1840, Giulio (70) also gave the true expression for the attraction of a spherical segment, and noted that several other
corrections must be made in Carlini's calculations; the height
of the segment is 1.05 miles, instead of 1 mile; the length of
the pendulum as determined by Biot must be corrected for an
error in the rule used to find the length, and for the altitude of
Bordeaux. Moreover, the reductions to vacuo had not been made
properly in either case. When all these corrections had been
applied to Carlini's results, Giulio found the value of Δ to be 4.95.

* This signification of Δ will be retained throughout the rest of the
volume.

Saigey (74, p. 155) makes the observed pendulum length corrected to vacuo 993.756 mm., and the calculated length 993.617 mm. With these numbers the value of Δ becomes 6.15.

Zanotti-Bianco (148½, pt. 2, p. 136) mentions that Knopf (149½) has compared the value of gravity as observed by Carlini on the top of the mountain with the value calculated for the same place from observations made on the same parallel of latitude, and found for Δ. 5.08.

AIRY, WHEWELL AND SHEEPSHANKS AT DOLCOATH MINE.

In 1826, Drobisch, in an appendix to a pamphlet on the figure of the moon (57), suggested that experiments be made on the change in the period of a pendulum when carried from the surface of the earth to the bottom of a mine; he gave the theory of the experiments and calculated the change resulting from certain hypotheses. It is interesting to recall the fact that Bacon proposed the same investigation two centuries earlier. (See p. 1.)

At the very same time, unknown to Drobisch, experiments of this nature were being tried in England by Airy and Whewell at the copper mine of Dolcoath in Cornwall. Their method was to swing one invariable pendulum at the mouth of the pit and compare its rate, by Kater's method of coincidences, with that of a standard clock, and at the same time perform the same operation upon another pendulum and another clock at a depth of 1220 ft. in the mine. The pendulums were then exchanged and the operations repeated. The greatest difficulty experienced was that of comparing the rates of the two clocks. The first series of experiments was abruptly stopped on account of the damage received from fire by the lower pendulum. A short account of the method was published in 1827 (60), and Drobisch translated it for Poggendorff's *Annalen* (59), wherein he gives also a more complete account of the theory and an application of his equations to Airy's observations. Assuming the mean density of the surface layer of the earth to be 2.587, the experiments gave about 20 for the value of Δ. Drobisch contends that the surface density should be taken to be 1.52, considering how large an amount of the surface layer is water.

Two years later Airy and Whewell, assisted by Mr. Sheepshanks and others, attempted to repeat the experiments ; but after overcoming various anomalies in the motions of the pendulums, the observations were stopped by a fall of rock in the mine. The value of Δ found from this series was about 6. A full account of the experiments was printed privately (62) in 1828, and Drobisch translated the pamphlet for the *Annalen* (63).

REICH'S FIRST EXPERIMENT. In 1838, F. Reich, Professor of Physics in the Bergakademie at Freiberg, published in book form (67) the account of a series of experiments carried on by him since 1835 to find, after the method of Cavendish, the mean density of the earth. The adoption of the mirror and scale method of measuring deflections seemed to him to promise a means of overcoming many of the difficulties against which Cavendish had contended. The final observations were made in the year 1837.

In order to avoid the effects due to irregularities of temperature, the apparatus was set up in a cellar room which was carefully closed up, and the observations made through a hole in the door. The arm of the balance was 2.019 m. long, and its moment of inertia was found after the manner used by Gauss for a magnet. The average weight of each of the balls was 484.213 gr., and their distance below the arm was 77 cm. They were composed of an alloy of about 90 parts tin, 10 parts bismuth, and a little lead. The attracting masses were of lead 45 kg. in weight and about 20 cm. in diameter, and hence much smaller than those used by Cavendish. They were suspended from pulleys running on rails parallel to the arm of the balance, and could be quickly moved from the null to the attracting positions. Only one mass was in the attracting position at a time, on account of the fact that in every one of the four attracting positions the distance from the mass to the ball was slightly different; whereas Cavendish used both masses at once. The distance from mass to ball was measured at each observation by means of a telescope moving along a horizontal scale, and not once for all as was done by Cavendish.

After the suspended system was set up, Reich found a continual changing of the zero-point, which often lasted for 6 months. In his final observations this was not noticeable because of the length of time, $1\frac{1}{2}$ years, which intervened between the initial and final experiments. Accordingly the second means of the elongations were found by him to be more constant than Cavendish found them. The following table will show this, and also illustrate Reich's method of finding the time of vibration :

Extremes	1st mean	2d mean	Time of passage	
			at 74 5	at 75 5
95.2	76.00		IVʰ 27′ 36″.8	IVʰ 27′ 30″.8
56.8				
	73.85	74 925	34 21 .2	34 29 .2
90.9				
	75.90	74.875	41 22 .4	41 12 .4
60.9				
	74.25	75.075	47 59 .2	48 10 .8
87.6				

Average or 3d mean 74.9583
Time of vibration determined from passage of 74 5

$$\left. \begin{array}{l} \text{IV}^h \ 41' \ 22''.4 - \text{IV}^h \ 27' \ 36''.8 = 13' \ 45''.6 \\ 47 \ 59 \ .2 - \quad\quad 34 \ 21 \ .2 = 13 \ 38 \ .0 \end{array} \right\} \text{aver. } 13' \ 41''.8$$

Time of vibration determined from passage of 75 5

$$\left. \begin{array}{l} \text{IV}^h \ 41' \ 12''.4 - \text{IV}^h \ 27' \ 30'' \ 8 = 13' \ 41''.6 \\ 48 \ 10 \ .8 - \quad\quad 34 \ 29 \ 2 = 13 \ 41 \ .6 \end{array} \right\} \text{aver. } 13' \ 41'' \ 6$$

By interpolation the time of a double vibration across 74.9583 is 13′41″.708. This differs somewhat from Cavendish's method, as will be seen by a reference to page 65. Reich considered it a more accurate method than that of Cavendish, and remarks that when he applied the latter's method to the above observations he got results not very different, but on applying his own method to Cavendish's observations he got somewhat different results. Reich's method was adopted by Baily (79, pp. 44 and 47) in his experiments.

When in the above way at least three passages across the two median points had been observed, Reich waited until the arm was in its next extreme position and seemingly at rest and then rapidly moved the attracting mass to its new position. It was always so moved as to increase the swing. He assumed that the motion was instantaneous, and used the last extreme of the one series as the first of the next. Baily (79, p. 46) followed him in this departure from the procedure of Cavendish. Cornu and Baille (142) have pointed out that this method leads to error, and we shall refer to the matter again when describing the results obtained by Baily (page 116). From each such set of 4 extremes the resting-point and the time of vibration are found. From 2 such sets the deviation could be found, and the mean density of the earth calculated. Reich did not, however, proceed in that way, but deduced one value of Δ from all the observations of each day ; that is, he took the average of all the deviations of that day for the final mean deviation, and the average of all the times of vibration for the final mean time of vibration, and from these deduced one value for the mean density of the earth.

In applying corrections to the equations derived from a simplified form of the theory of the experiment, Reich followed Cavendish exactly. 57 observations were made, from which 14 determinations of the value of Δ were deduced. The mean of all, when corrected for the centrifugal force, was 5.44±.0233, a result almost coinciding with Cavendish's. Reich admits at the end of his paper that there were certain anomalies in the motion of the beam which he could not account for.

A second series of 6 observations with iron masses 30 kg. in weight and 20 cm. in diameter gave for Δ, 5.4522, which proves that no disturbance could have arisen from magnetic action.

Valuable concise accounts of Reich's experiment are given by Beaumont (66), Baily (68, 69 and 79, pp. 96–8), Schell (135), Poynting (185, pp. 48–50) and Fresdorf (186½, pp. 20–2).

BAILY. While Reich was making the investigations just referred to, a very comprehensive and elaborate series of experiments upon almost the same plan was being carried on by the English astronomer F. Baily. These experiments were undertaken at the instance of the Royal Astronomical Society, and in aid of them a grant of £500 was made by the British government. The results were published in 1843 (79). They were carried on in one of the rooms of Baily's residence, a one-story house standing detached in a large garden. The apparatus was almost the counterpart of that of Cavendish, except that the balls were not suspended from the balance arm, but were screwed directly on to its ends. The balance and its mahogany case were, moreover, suspended from the ceiling, and the attracting masses rested on the ends of a plank movable on a pillar rising from the floor. As a protection against changes of temperature this apparatus was then surrounded by a wooden enclosure. The masses were of lead rather more that 12 in. in diameter, weighing 380.469 lbs. each. Torsion rods of deal and of brass, each about 77 in. long, were employed, and their motion was observed by the mirror and scale method. Balls of different materials and of various diameters were experimented upon : viz., 1.5 in. platinum, 2 in. lead, 2 in. zinc, 2 in. glass, 2 in. ivory, 2.5 in. lead, and 2.5 in. hollow brass. The mode of suspension was varied greatly, both single and double suspension wires being used, and the material and distance apart of the bifilar wires being frequently changed. The length of the suspending wires was ordinarily about 60 in., and the time of vibration varied from about 100 to 580 seconds.

The experiments were begun in Oct. 1838, and carried on for 18 months, until about 1300 observations had been made ; when, on account of the great discordance of the results, a stop was made. Prof. Forbes suggested that these anomalies might arise from radiation of heat, and advised the use of gilt balls and a gilt case. These changes were made, and the torsion box also lined with thick flannel. They turned out to be decided improvements, although some anomalies still existed, and it is evident that the choice of a place for setting up the apparatus was not a good one.

Baily adopted the method of Reich for reducing the time required to make the number of turning-points requisite for calculating the deviation and period ; that is, the masses were moved quickly from one near position to the other, and the last turning-point of one series served for the first of the next. Three new turning-points were observed at each position of the masses, and each group of 4 was called an " experiment." 2153 such experiments were made during the years 1841–2. The time of vibration was found for each experiment after the method adopted by Reich. In deducing the mean density of the earth from the observations Baily proceeded quite differently from Reich. There was always a slow motion of the zeropoint, and Baily, in order to take account of this, combined the deflections and periods *in threes*. The difference between the deflection of the 2d experiment and the average of the 1st and 3d is twice the mean deviation. The average of the period of the 2d experiment with the average of the 1st and 3d is the mean period. From the mean deviation and mean period so found a value of Δ is deduced. Another was then found from comparing the 3d experiment with the 2d and 4th, and so on. The mean of all the experiments gave for Δ, $5.6747 \pm .0038$. Some of the experiments were made with the brass rod alone, without any balls, the mean result for which was $5.6666 \pm .0038$.

The mathematical analysis of the problem was given by Airy, and is incorporated in Baily's paper (79, pp. 99–111); it is also to be found in Routh's *Rigid Dynamics*, 1882, pt. 1, pp. 359–64.

Baily published a condensed account of his work in several journals (75, 76, 77, 78 and 80). A careful discussion of it is given by Schell (135) and by Poynting (185, pp. 52–7).

In 1842, Saigey (74) wrote a full account of all the experiments made before that date ; he gives his reasons for considering the pendulum method of finding Δ the least accurate, the mountain method somewhat better, and the torsion method the best. He finds great fault with the work of Baily, and considers that his results are not so worthy of confidence as those of Cavendish. Saigey contends that the anomalies observed by Cavendish, Reich, and Baily cannot be accounted for by radiation of heat, as Forbes suggested, because the balance swings in an enclosure all points of which are at the same temperature (thus begging the question); he confidently remarks that these anomalies are caused by the passage of air into or out of the case as the barometric pressure changes. The values of Δ found by Baily increased from 5.61 to 5.77 as the density of the balls used changed from 21.0 to 1.9 respectively; Saigey thinks that this must arise from an error in calculating the moment of inertia of the balance arm. He devises a graphical method of making proper allowance for this supposed error, and deduces as the final mean of all the experiments of Baily a value 5.52, the extremes being 5.49 and 5.55.

Saigey made a new determination of Δ (74, vol. 12, p. 377), from the difference $6''.86$ of the astronomical and geodetical latitudes of Évaux as calculated by Puissant. Applying the method used by Hutton for Schehallien, and later by James and Clarke for Arthur's Seat, he found the ratio of Δ to the surface density of France to be 1.7. Assuming the latter density to be 2.5, the former becomes 4.25.

In 1847, Hearn tried to account for the anomalies in Baily's results by assuming a magnetic action. He worked out the theory (81) of such action, and found that it must be of a very fluctuating nature and may be either positive or negative, and even greater in magnitude than the force of gravitation. That such a magnetic action does not really exist is to be deduced from Reich's results with iron masses (see page 116).

Montigny offered to the Royal Academy of Belgium, in 1852, a memoir in which he attributed the peculiarities in the behaviour of the torsion pendulum in the experiments of Cavendish and of Baily to the rotation of the earth. Schaar (85), to whom the memoir was referred by the Society, proved that the rotation of the earth could not produce such effects, and the memoir was not published.

It was Cornu and Baille who first pointed out (142), in 1878, the main error in Baily's method. It lies in his taking the 4th reading of the turning-point of one series of experiments as the 1st of the next, as already explained. They shewed that the rotation of the plank holding the masses could not be performed rapidly enough to get the masses into the new position before the arm had begun its return journey. They therefore rejected the 1st of each series of 4 readings, and calculated Δ from the other 3 in 10 cases taken at random from some of Baily's most divergent values, and found 5.615 instead of 5.713. Reducing Baily's final value in the same proportion they got 5.55.

A curious relation between density and temperature as presented in Baily's determinations was pointed out by Hicks (166), in 1886. The mean density seems to fall with rise of temperature. The most probable explanation of this is given by Poynting (185, p. 56), who remarks that the experiments with the light balls happened to be made in winter, and those with the heavy balls in summer. Hicks also refers to several slight corrections to be made in Airy's discussion of the theory —viz., for the air displaced by the attracting masses, for the inertia of the air in which the balls move, and for expansion with change of temperature.

REICH'S SECOND EXPERIMENT. Ten years after the appearance of Baily's memoir, Reich published (83) an account of some further experiments with his apparatus. In the beginning of his paper he pointed out that Baily's method of combining the results of the separate experiments was better than that used by himself. He proceeded to calculate the results of his first experiments by Baily's method and found for Δ the value $5.49 \pm .020$.

Being impressed with the anomalies in Baily's observations, and especially with the variation of the final results with the density of the balls, Reich determined to repeat his experiments. His apparatus was set up this time in a second-story room, and Baily's devices were employed in order to avoid the effects of temperature changes. The only important change in the arrangement of the apparatus was in the placing of the large mass. It was now set in one of four depressions 90° apart in a circular table revolving under the balance about a vertical axis passing through the centre of one of the balls; thus no correction was necessary for the attraction of the table and its supports upon the ball. The balls and masses were those used in the first experiment. Three series of experiments were made during the years 1847–50, one with a suspending wire of thin copper, one with thick copper, and one with a bifilar iron suspension. The final mean density of the earth was found to be 5.5832 ± .0149.

In order to make a test of Hearn's explanation (see page 118) of the peculiarities in Baily's results, Reich made some further experiments. He kept the North pole of a strong magnet near the attracting lead mass for a whole day, and then suddenly rotated the mass through 180° about a vertical axis; but no effect was evident. Hence variations in the result are not due to the magnetizing of the masses by the earth, or similar causes. He then took off the tin balls and substituted successively balls of bismuth and of iron. The values of Δ were respectively 5.5233 and 5.6887; the largeness of the latter denotes possibly a diamagnetic action of the lead mass; but it shews that under the original circumstances no measurable effect could have arisen from magnetic action.

Prof. Forbes had suggested * to Reich that Δ could be found from the period of the balance only, by noting the variation of the time of vibration with the position of the attracting masses. Reich made some experiments of this nature by placing two lead masses diametrically opposite to each other, first so that the line joining them was perpendicular to the vertical plane through the torsion arm, and next was in the plane. This caused no deviation, but only a change in the time of swing of

* We have seen (page 106) that this method was suggested earlier, in Gehler's *Physikalische Worterbuch*, and the equations given by Brandes (42).

the balance. The value of Δ found in this way was 6.25, but the apparatus was not well devised for the work.

Several abstracts of Reich's paper are to be found (84, 86, 87 and 185, pp. 50–2).

AIRY'S HARTON COLLIERY EXPERIMENT. We have already referred to Airy's experiments in the Dolcoath mine in 1826–8. In 1854, he again undertook to carry out investigations (100) along the same lines, the introduction of the telegraph having made easy the comparison of the clocks at the top and bottom of the mine. He selected the Harton Colliery, near South Shields, for the experiments, which were carried out by six experienced assistants of whom Mr. Dunkin was the chief. The two stations were vertically above each other and 1256 ft. apart. The apparatus was the best obtainable, and special precautions were taken in order that the pendulum supports might be rigid.

Simultaneous observations of the two pendulums were kept up night and day for a week ; then the pendulums were exchanged and observations taken for another week. Two more exchanges were made, but the observations for them both were made in one week. Each pendulum had six swings of nearly 4 hours each on every day of observation, and between successive swings the clock rates were compared by telegraphic signals given every 15 seconds by a journeyman clock.

The corrections and reductions were carried out by Airy in a very elaborate manner. The results of the 1st and 3d series agree very closely, as do those of the 2d and 4th, showing that the pendulums had undergone no sensible change. By comparing the mean of the 1st and 3d series with the mean of the 2d and 4th, the ratio of the pendulum rates at the upper and lower stations is obtained independently of the pendulums employed. The final result gave gravity at the lower station greater than gravity at the upper by $\frac{1}{19\,286}$th part, with an uncertainty of $\frac{1}{2\,10}$th part of the increase : or the acceleration of the seconds-pendulum below is 2″.24 per day, with an uncertainty of less than 0″.01.

In order to calculate what this difference should be, suppose the earth to be a sphere of radius r and mean density Δ, surrounded by a spherical shell of thickness h and density δ, then a simple analysis shews that $\dfrac{\text{gravity below}}{\text{gravity above}} = 1 + \dfrac{2h}{r} - \dfrac{3h\delta}{r\Delta}$ (compare p. 31). Airy gives a discussion of the effect of surface irregularities ; it is shewn that, supposing the surface of the earth near the mine to have no irregularities, the effect of those at distant parts of the earth may be neglected. He also assumes that there is no sudden change of density just under the mine. He proves that the effect of a plane of 3 miles in radius and of the thickness of the shell is $\frac{23}{24}$ of that of the whole shell, so that only the neighbouring country need be surveyed. Since the upper station is only 74 ft. above high water, it will be sufficient to assume that any excess or defect of matter exists actually on the surface. A careful survey of the environs of the mine was made, and allowance made for each elevation and depression. The general result is that the attraction of the regular shell of matter is to be diminished by about $\frac{1}{280}$th part ; $\dfrac{\text{gravity below}}{\text{gravity above}}$

$=1.00012032-.00017984\times\dfrac{\delta}{\Delta}.$ Now from the pendulum experiments Airy found $\dfrac{\text{gravity below}}{\text{gravity above}}=1.00005185\pm.00000019$;

hence $\dfrac{\Delta}{\delta}=2.6266\pm.0073.$ Prof. W. H. Miller found the average density of the rocks in the mine to be 2.50; hence $\Delta=6.566$ $\pm.0182.$

Airy had intended that the temperatures at the two stations should be the same, but the temperature of the lower station was 7°.13 F. higher than that of the upper. In a supplementary paper (101) Airy makes a correction for this temperature difference in two distinct ways, giving for the corrected Δ, 6.809 and 6.623 respectively. In this paper Stokes (102) investigates the effect of the earth's rotation and ellipticity in modifying the results of the Harton experiments. It was found to be small, changing Δ from 6.566 to 6.565.

Airy published several preliminary notices of his work (88, 89 and 122), abstracts of which appeared in several journals (90, 91, 92, 98 and 111). Valuable résumés of the main paper are also to be found (105, 107, 109, 112 and 119).

Haughton (106, 110, 113 and 116) gave a rough but simple method of deducing Δ from Airy's figures, and arrived at 5.48 as the value of Δ. Knopf (149½) has severely criticized this calculation. Another simple formula for the same purpose was given by an anonymous writer (114). On the effect of great changes in density below the under station one should read the paper by Jacob (118 and 121) already referred to. Scheffler (134) published in 1865, though it is dated 1856, the proposal of an experiment similar to Airy's, but made no reference to any earlier proposals of the same kind. Folie (136) calculated, in 1872, the attraction at the two stations in a manner different from Airy's, by considering the shell as made up of 2 parts. Using Airy's data he arrived at 6.439 as the value of Δ. Valuable summaries and criticisms of Airy's work are given by Schell (135). Zanotti-Bianco (148½, pt. 2, pp. 146–60), Poynting (185, pp. 24–9) and Fresdorf (186½, pp. 13–7).

JAMES AND CLARKE. As a result of the calculations made from the observations taken for the Ordnance Survey of Great Britain and Ireland (104, 117, 120, 124, 125 and 126) by Lt. Col. James, it was found that the plumb-line was considerably deflected at several of the principal trigonometrical stations. It was evident from the nature of the ground at the places under consideration, that this deflection was due to irregularities of the surface. In order to study this action more carefully James decided to have the Schehallien experiment repeated at Arthur's Seat, near Edinburgh (103 and 125, pp. 572–624). The observations were made during Sept. and Oct., 1855, with Airy's zenith-sector, on the summit of Arthur's Seat (A), and at points near the meridian on the north (N) and south (S) of that mountain, at about one-third of its altitude above the surrounding country. After corrections had been applied, the results were as follows:

Station	Astronomical lat.$\equiv$A	Geodetical lat.$\equiv$G	A$-$G
S	55° 56′ 26″.69	55° 56′ 24″.25	2″.44
A	56 43 .69	56 38 .44	5 .25
N	57 9 .22	57 2 .71	6 .51

It will be noticed that even on the summit of the hill there is an attraction of more than 5″ toward the south, which can not be due to the hill. Similarly, to the south of the hill the attraction is not toward the north as we might expect. It is evident that there is present some other attracting force, besides that of Arthur's Seat, which appears to produce a general deflection of 5″ toward the south.

Capt. Clarke, who made all the calculations, in order to find the attraction according to Newton's law, used a modification of the method of Hutton. He took account of all the surface irregularities within a radius of about 24000 ft. The resulting value for the ratio of the density of the rock composing the hill to that of the whole earth was .5173±.0053. James investigated the density of the rocks of Arthur's Seat and found it to be on the average 2.75. This gives for Δ the value 5.316 ±.054.

In order to see whether the general deflection of 5″ could be accounted for by the presence of the hollow of the River Forth to the north and the high land of the Pentland Hills to the south, Clarke extended the calculated attraction to the borders of Edinburghshire, some 13 miles away. He was able in this way to account for a general deflection of 2″.52, and he thought that by carrying the calculations to Peeblesshire the whole 5″ might be accounted for.

Several abstracts of the original paper have been published (108, 115 and 123). Poynting (185, pp. 19–22) has given a valuable criticism of the work.

In connection with this investigation might be mentioned the various writings on the subject of local attractions. Any one wishing to become acquainted with this subject should read Airy's account of his "flotation theory" (94 and 97), Faye's account of his "compensation theory" (130, 146½ and 147), Pratt's papers (93, 96 and 99), Saigey (74), Struve (129), Pechmann (131), the treatises of Pratt (133), Clarke (149) and Helmert (148, vol. 2). Many other references to papers by these men as well as by Schubert, Peters, Keller, Bauernfeind and others are to be found in the Roy. Soc. Cat. of Scientific papers and in Gore's "A Bibliography of Geodesy" (174). See also note on page 31 and remarks on page 56. We might here recall the determination of Δ by Saigey from local attraction (see page 118). Pechmann (131) in the same way found in the Tyrol, in 1864, two different values for Δ, 6.1311±.1557 and 6.352± .726, having assumed the density of the earth's crust to be 2.75. We shall refer later on to the determinations of Mendenhall and Berget.

CORNU AND BAILLE. In 1873, Cornu and Baille published a short paper (137) stating that they had undertaken to repeat the Cavendish experiment under conditions as different as possible from those previously employed. They began by making a thorough study of the torsion-balance in order to learn under what conditions it would have the greatest precision and sensitiveness. They found among other things that the resistance of the air was proportional to the velocity (141, 142, 143 and 157).

The apparatus was set up in the cellar of the École Polytechnique. The arm of the balance was a small aluminium tube 50 cm. long, carrying on each end a copper ball 109 gr. in weight. The suspension wire was of annealed silver 4.15 m. long, and the time of vibration of the system 6′ 38″. The attracting mass was mercury which could be aspirated from one spherical iron vessel on one side of one of the copper balls to another vessel similarly situated on the other side of the ball. This method got rid of the disturbances arising from the movement of the lead masses in the Cavendish form of the experiment. The iron vessel was 12 cm. in diameter and the mercury weighed 12 kg. Another great improvement was the reduction of the dimensions of the apparatus to $\frac{1}{4}$ of that used by Cavendish, Reich and Baily, the time of oscillation and the sensitiveness remaining the same. The motion of the arm was registered electrically.

Two series of observations were made ; one in the summer of 1872 gave $\Delta = 5.56$, and the other in the following winter 5.50. The difference was explained by a flexure of the torsion-rod, and the former result was considered the better.

In a later report (142) they refer to some changes made in their apparatus ; they increased the force of attraction by using 4 iron receivers, 2 on each side of each copper ball, and they reduced the distance between the attracting bodies in the ratio of $\sqrt{2}$ to 1. The time of vibration, 408″, remained the same within a few tenths of a second for more than a year. The new value of Δ was 5.56.

We have already referred (page 119) to the fact that Cornu and Baille found out the error in the Baily experiments.

A final account of these experiments has not yet been published. Abstracts of the papers cited are given by Poynting (185, pp. 57–8) and by several journals (138 and 139).

JOLLY. In 1878, von Jolly of Munich published an account (144 and 145) of the results of his study of the beam balance as an instrument for measuring gravitational attractions. He discussed the sources of error in the balance readings and methods of eliminating them. The variations due to temperature effects are very difficult to avoid, but by working in the mornings only, and by covering the balance case with another lined inside and out with silver paper, it was found to be possible to get quite concordant results.

Jolly applied the balance to test the Newtonian law of the distance. Two extra scale pans were suspended by wires from the ordinary scale pans of the balance and 5.29 m. below them. The wires and lower scale pans were enclosed to prevent oscillations from air currents. Two kilogramme masses of polished nickel-plated brass were balanced against each other, first both in the upper scale pans, and then one in the upper and the other in the lower pan, in each case double weighings being made after the manner of Gauss. The motion of the beam was noted by the mirror and scale method, the mirror being fixed at the middle of the beam and perpendicular to its length. If r is the radius of the earth at sea-level, and h a height above it, then a

mass Q_1 at sea-level weighs Q_2 at h, where $Q_2 = Q_1 \left(1 - \dfrac{2h}{r}\right)$ approximately. Jolly found by experiment $\dfrac{Q_2}{Q_1} = \dfrac{1\ 000\ 000 - 1.5099}{1\ 000\ 000}$,

whereas the equation gives $\dfrac{Q_2}{Q_1} = \dfrac{1\ 000\ 000 - 1.662}{1\ 000\ 000}$. The difference, .152 mg., Jolly thought, was due to local attractions. He proposed to repeat the experiment at the top of a high tower, and at the same time to find the mass of the earth by noting the change in weight of one of the masses in the balance when a large lead ball was brought beneath it.

The results of these experiments (153 and 154) were published in 1881. The distance between the scale pans was now 21.005 m. The arm of the balance was 60 cm. long, and the maximum load 5 kg. Four hollow glass spheres of the same size were made and in each of two 5 kg. of mercury were put, and all were sealed up. Each scale pan had always one sphere in it, and thus air corrections were avoided. An observation was made as follows : first the mercury-filled spheres were balanced in the upper pans, and then one in the upper pan was balanced against the other in the lower. The change in weight observed was 31.686 mg.; whereas the change as calculated from the formula should have been 33.059* mg. The difference is in the same direction as in the earlier experiment.

A sphere of radius .4975 m. and weight 5775.2 kg. was then built up out of lead bars under the lower scale pan which received the mercury-filled globe. The distance from the centre of this sphere to that of the globe was then .5686 m. The attraction of the sphere for the mercury-filled globe when in the upper pan was neglected.

Observations were made exactly as before, and the change in weight was 32.275 mg. The increase in weight due to the presence of the lead is therefore .589 mg. Knowing the density of the lead to be 11.186, a simple calculation gives for the mean density of the earth $5.692 \pm .068$.

An account of these experiments is given by Helmert (148, vol. 2, pp. 380-2), Zanotti-Bianco (148½, vol. 2, pp. 175-82), Wallentin (154½), Keller (167), Poynting (185, pp. 61-4) and Fresdorf (186½, pp. 23-5).

MENDENHALL. In 1880, Prof. T. C. Mendenhall described (150) a method of finding the period of a pendulum such that a determination required 20 or 30 minutes only. At the beginning and end of this time the pendulum throws a light trip-hammer of wire which breaks a circuit and makes a record on a chronograph on which a break-circuit clock is also marking. The advantage of such an arrangement, in addition to the short time required, is that the arc of vibration may be small and will change very little. Mendenhall expressed a determination to find the variation of the acceleration due to gravity on going from Tokio to the top of Mount Fujiyama.

A year later the results of these experiments were published (151), having been made in Aug., 1880. An invariable pendulum was used, made from a Kater's pendulum by removing one ball and knife-edge. Its period at Tokio (barometer 30 in. and temperature 23°.5 C.) was .999834 sec. On the top of Fujiyama the barometer stood nearly stationary at 19.5 in. during the observations, and the thermometer at 8°.5. After approximate corrections were made for buoyancy, the time, reduced to Tokio conditions, was 1.000336 sec. Assuming g at

* According to Helmert this should be 33.108 and according to Zanotti-Bianco 33.053.

Tokio to be 9.7984, as he had found in the previous year, it follows that at the summit of Fujiyama it is 9.7886.

No exact triangulation of the region had been made, but Mendenhall assumed Fujiyama to be a cone 2.35 miles high standing on a plain of considerable extent. The angle of the cone was measured from photographs and found to be 138°. Fujiyama is an extinct volcano, said to have been made in a single night, and hence its composition ought to be homogeneous. Its average density was taken as 2.12, but no great reliance can be placed on this number. Corrected for the difference in latitude, 19', between Tokio and Fujiyama, the time at its base, supposing the hill taken away, would be .999847 sec. The density of the earth, calculated from these data after the manner of Carlini, was found to be 5.77.

Fresdorf (186½, pp. 11-13) describes fully the experiments and points out an error in Mendenhall's calculations ; the corrected value for Δ is 5.667. Poynting (185, pp. 39-40) gives an abstract of the papers referred to.

STERNECK. Major von Sterneck has made several investigations of the variation of gravity beneath the earth's surface. The earliest experiments (155) were made, in 1882, in the Adalbert shaft of the silver mine at Príbram in Bohemia. The method employed was to carry an invariable half-second pendulum and a comparison clock from one station to another, and find the period by the method of coincidences, the clock being compared with a standard clock by carrying a pocket chronometer from one to the other. The pendulum, of brass, was a rod 24 cm. in length carrying a lens-shaped bob weighing 1 kg. The knife was of steel whose edge was so cut away that it rested on a glass plate on two points only. The apparatus was always enclosed in a glass case to prevent air currents. The 3 stations at which observations were made were at the surface, 516.0 m. and 972.5 m. below the surface respectively. The respective periods at these stations were .5008550, .5008410 and .5008415 seconds, and the resulting values of Δ, found from Airy's formula, were 6.28 and 5.01, the density of the surface layer being taken as 2.75. It will be noticed that the values of g at the two underground stations are practically the same, and the results are unsatisfactory.

A year later (156) von Sterneck repeated his experiments at

the same stations and at two additional ones. In order that his observations might be independent of the rates of the clocks used in finding the periods, Sterneck introduced an important modification of the method adopted by Airy and by himself in his earlier investigations. He made another pendulum similar to the one described above ; one of these was always at the surface station and the other at one of the underground stations, and their relative periods were compared by means of electric signals sent simultaneously from a single clock. This clock kept a circuit closed for half a second every other half second and operated a relay with a strong current at each station. The passage of the "tail" of the pendulum in front of a scale was observed by means of a telescope, in the focal plane of which was a shutter moved by the relay current every half second, and at those instants only was the picture of the tail of the pendulum allowed to pass to the eye through the telescope. The time of a coincidence was when at one of these flashes the tail appeared exactly at the middle of the scale ; the time between two successive coincidences determines the period of the pendulum. The observer at each of the two stations is thus finding the period of his pendulum in terms of exactly the same unit of time. When the observations were corrected, it was found that the period at the highest underground station was less than that at the next lower station, and the determination at the former station was consequently not used. The values of Δ as determined from observations at the other stations were 5.71, 5.81 and 5.80, with a mean of 5.77. Helmert (148, vol. 2, p. 499) has made a recalculation and finds that these numbers should be 5.54, 5.71, 5.80 and 5.71 respectively.

Von Sterneck used his results at the surface and at these underground stations to express g as a function of the depth. Calling the value of g at the surface unity, and measuring r from the centre of the earth and calling it equal to unity at the surface, he deduced the following expression for the value of g at any depth :

$$g = 2.6950\,r - 1.8087\,r^2 + .1182\,r^3.$$

This would make g a maximum, 1.06, at $r = .78$. The density would be expressed by the formula $d = 15.136 - 12.512\,r$, giving 15.136 for its value at the centre of the earth, and 2.624 at the surface. These relations are at least suggestive if not convincing.

During the year 1883 von Sterneck used the same method and apparatus to determine the variation in gravity for 3 stations above the earth's surface at Kronstadt. He found (158) gravity greater at a higher point (Schlossberg) than at a lower (Zwinger), and proved that neither the formula of Young (see page 31) nor that of Faye and Ferrel for the reduction to sea-level gave satisfactory results.

Twice in this year Sterneck made investigations at Krušná hora in Bohemia. Here there was a mine with a horizontal gallery 1000 m. long, and he wished to find the effect of the overlying sheet of earth upon the value of gravity at various points in the gallery. The same apparatus was used after some improvements had been made. Observations were taken at the mine mouth and at points 390 and 780 m. from the mouth, and 62 and 100 m. respectively below the surface of the ground. The results shewed that gravity in the plateau increased with the depth of the super-incumbent layer by the half of the amount by which it would have changed in free space when the distance from the centre of the earth was changed by the same amount. Observations were made at 4 stations above ground also at different elevations, and it was found that the Faye-Ferrel rule accounted for the differences between them much better than did the Bouguer-Young rule.

Further experiments (164) were made, in 1884, at Sághegy in Hungary, and elsewhere, with results similar to those described above. An important improvement was made in the method of observing the coincidences. They were now observed by the reflections of an electric spark from two mirrors, one fixed on the pendulum stand, and the other attached to the pendulum and when at rest parallel to the first. The spark was made by the relay circuit every half second.

In 1885, Sterneck made a series of observations (165) at the mouth and at 4 underground stations in the Himmelfahrt-Fundgrube silver mine at Freiberg in Saxony. He was led to do so by the publication of the results of some pendulum measurements made there, in 1871, by Dr. C. Bruhns, who had found that gravity decreased with the depth. Using Airy's formula, von Sterneck found the following values for Δ at the 4 underground stations in the order of their depth : 5.66, 6.66, 7.15 and 7.60, the density of the mine strata being 2.69. These results indicate an abnormal increase of gravity with depth. Von

THE LAWS OF GRAVITATION

Sterneck noticed that in these experiments, as well as in those made at Přibram, the increase in gravity is nearly proportional to the increase in temperature. But although Hicks (166), as we have seen (page 119), discovered a connection between the values of Δ and the temperatures in Baily's experiments, and Cornu and Baille (page 125) got a larger result for Δ in summer than in winter, we have no reason for looking upon the variations in temperature as an explanation of the anomalies under consideration. An interesting criticism of von Sterneck's work is given by Poynting (185, pp. 29–39). Short accounts of it are given by Fresdorf (186½, pp. 17–9) and Günther (196½, vol. 1, p. 189).

WILSING. In 1887, J. Wilsing (170) made at Potsdam a determination of the mean density of the earth by means of an instrument which is called the pendulum balance, and is the common beam balance turned through 90°. It is practically a pendulum made of a rod with balls at each end and a knife-edge placed just above the centre of gravity. The instrument used by Wilsing consisted of a drawn brass tube 1 m. long, 4.15 cm. in diameter and .16 cm. thick, strengthened near the middle where the knife-edge is affixed. The knife-edge and the bed on which it rested were of agate, and 6 cm. long. To the ends were screwed the balls of brass weighing 540 gr. each, and on the upper ball was a pin carrying discs which were used for finding the moment of inertia and the position of the centre of gravity of the pendulum. Its motion was observed by the telescope and scale method, a mirror being attached to the side of the pendulum parallel to the knife-edge. The pendulum was mounted on a massive pier in the basement of the Astrophysical Observatory in Potsdam, and was protected from air currents by a cloth-lined wooden covering.

The attracting masses were cast-iron cylinders each weighing 325 kg. They were so arranged on a continuous string passing over pulleys that when one was opposite the lower brass ball on one side of the pendulum the other was opposite the upper ball on the other side. Their relative positions could be quickly changed from without the room, so that the former mass came opposite the upper ball and the latter mass opposite the lower ; the deflection was now in the opposite direction from what it was in the first case.

The double deflection due to the change in position of the masses, and the time of vibration are the quantities required for the determination of Δ. The readings for these quantities were made by the method of Baily, which has been already described. The time of vibration was determined first with the discs on top of the upper ball, then with one removed and then with still another removed. In this way the moment of inertia was obtained. The theory of the instrument is complicated, and for it reference must be made to the original paper. The result obtained for Δ was $5.594 \pm .032$.

In 1889, Wilsing published (172) an account of some further observations with the same apparatus, some slight changes having been made in it in the meantime. Extra precautions were taken in order to avoid the effects of variations of temperature. Experiments were made with the old balls, with new lead balls, and with the pendulum rod alone. The mean result from these was $5.588 \pm .013$; and the final average of all his determinations $5.579 \pm .012$.

A preliminary paper (163) was read by Wilsing before the Berlin Academy, and also an extract (169) of his first paper. A condensed translation of both papers was made by Prof. J. H. Gore (171) for the Smithsonian Report for 1888, and a short account of the work is given by Poynting (185, pp. 65–9) and by Fresdorf (186½, p. 28).

POYNTING. Prof. J. H. Poynting published in 1878 the results (146) of a study of the beam balance. He found that the sources of error were temperature changes producing convection currents and unequal expansion of the arms, and the necessity of frequently raising the knife-edges from the planes. He tried to overcome the former difficulty by taking the same precautions as those employed by users of the torsion balance ; and he did away altogether with the raising of the beam between weighings, and when the weights had to be exchanged held the pan fixed in a clamp.

The paper gives a description of his balance and illustrates how it can be used, (1) to compare two weights, and (2) to find the mean density of the earth. The motion of the beam was observed by means of a telescope and scale, the mirror being fixed at the centre of the beam. The deflection of the ray could be multiplied by repeated reflections between this mirror and another which was fixed and nearly parallel to the former. The centre of oscillation was determined after the method of Baily with the torsion balance. As a result of 11 observations Prof. Poynting found the mean density of the earth to be $5.69 \pm .15$. He felt justified, therefore, in proceeding to have a more suitable balance constructed in order to make a more careful determination of this quantity.

The investigation continued through many years, and the results (180 and 185, pp. 71–156) were not published until 1891. Many unforeseen difficulties arose during the progress of the work, but by patience and skill Poynting was able to overcome these difficulties and to begin to take observations in 1890. The balance was of the large bullion type, 123 cm. long, and made with extra rigidity by Oertling. It was set up in a basement room at Mason College, Birmingham. The principle upon which the experiment is based is as follows : two balls of about the same mass are suspended from the two arms of the balance. Beneath the balance is a turn-table carrying a heavy spherical mass vertically under one of the balls. The position of the beam is observed and the turn-table moved until the mass is under the other ball and the position of the beam again observed. The deflection measures twice the attraction of the mass for the ball. The attraction of the mass for the beam and wires, etc., is then eliminated by repeating these observations with the balls suspended at a different distance below the arm, for then the attraction of the mass on the balance remains the same, and we find the change in the attraction of the mass for the ball with change of distance. The calculation was complicated by the presence on the turn-table of another mass as a counterpoise to the former one ; it was smaller than that one and at a correspondingly greater distance from the centre. It was used because certain anomalies could be accounted for only on the supposition that the floor tilted when the turn-table was rotated with the large mass only upon it.

Instead of the ordinary mirror fixed on the beam, Poynting used the double-suspension mirror (see Darwin B. A. Rep., 1881). The riders were manipulated by mechanism from without, and the observer was stationed in the room above, whence he could make all changes and observations without opening the balance room. The attracted and attracting masses were made of an alloy of lead and antimony. The balls were gilded and weighed over 21 000 gr. each. The large mass weighed 150 000 gr., and the counterpoise about half as much. A first set of observations gave $\Delta = 5.52$. The attracting bodies were then all inverted in order to eliminate the effects of want of symmetry in the position of the turn-table, and of homogeneity in the masses. A new set of observations gave $\Delta = 5.46$. The difference between the results of the two sets must have been caused by a cavity or irregular distribution of density in the large mass, and by other experiments Prof. Poynting found that its centre of gravity was not at its centre of figure, but was nearly at the place at which his gravitational experiments would have suggested it to be. The mean result for Δ is taken to be 5.4934, and for the gravitation constant, G, 6.6984×10^{-8}.

Poynting remarks that the effects of convection currents are greater in the beam balance than in the torsion balance, since the motion of the former is in a vertical plane. He thinks that a balance of greatly reduced dimensions would have been preferable. The admirable way in which Prof. Poynting has utilized the common balance for absolute measurements of force caused the University of Cambridge to award him the Adams Prize in 1893.

For a short account of this work see Wallentin ($154\frac{1}{2}$).

BERGET. In 1756, Bouguer read before the Academy of Sciences the results (9) of some experiments made by him to determine whether the plumb-line was affected by the tidal motion of the ocean. He was not able to detect any such effect. Towards the middle of last century Boscovitch proposed (140, vol. 1, pp. 314 and 327) to place a long pendulum in a very high tower by the edge of the sea, where the height of the tide is very great, and to observe the deviation due to the rise of the water, and thence to calculate the mean density of the earth. Von Zach suggested (49, vol. 1, p. 17) a modification of the experiment. Boscovitch also proposed the use of a reservoir after the manner about to be described, used by Berget. In 1804, Robison, in his "Mechanical Philosophy" vol. 1, page 339, points out that a very sensible effect on the value of gravity might be observed at Annapolis, Nova Scotia, due to the very high tides there. The theory of this local influence is given in Thomson and Tait's "Natural Philosophy" pt. II., page 389. Struve (129) proposed to find Δ from observations on plumb-lines placed on each side of the Bristol Channel, and Keller (168) calculated the deflection of the plumb-line due to the draining of Lake Fucino. In 1893, M. Berget utilized this principle in order to find (181) the density of the earth.

He had the use of a lake of 32 hectares in area in the Commune of Habay-la-neuve in Belgian Luxembourg. The level of the lake could be lowered 1 m. in a few hours, and as quickly regained. He could thus introduce under his instrument a practically infinite plane of matter whose attraction could be calculated and observed. The apparatus used to measure the attraction was the hydrogen gravimeter such as Boussingault and Mascart (Comp. Rend., vol. 95, pp. 126-8) used to find the diurnal variation of gravity. The variation of the column of mercury was observed by the interference fringes in vacuo between the surface of the mercury and the bottom of the tube, which was worked optically plane. A first series of observations was made when the lake was lowered 50 cm., and another when it was lowered 1 m. A change of 1 m. caused a displacement of the mercury column of 1.26×10^{-6} cm. The value of the gravitation constant found was 6.80×10^{-8}, of Δ, 5.41 and of the mass of the earth 5.85×10^{27} gr.

M. Gouy remarks (182) that such a result would imply that the temperature remained constant during hours to $\frac{1}{5\,000\,000}$ of a degree, which is impossible. Pavillon, with the greatest care, was able to reach $\frac{1}{10\,000}$ of a degree only. So that the result given by Berget can not be so accurate as he supposed.

For a short account of the experiment see Fresdorf (186½, pp. 29-30).

Boys. Prof. C. V. Boys read before the Royal Society, in 1889, an important paper (175 and 176) on the best proportions and design for the torsion balance as an instrument for finding the gravitation constant. He shewed that the sensibility of the apparatus, if the period of oscillation is always the same, is independent of the linear dimensions of the apparatus ; and remarked that the statements of Cornu on this point (page 125) are not correct. There are great advantages to be gained by reducing the dimensions of the apparatus of Cavendish 50 or 100 times ; the main one is that the possibility of variation of temperature in the apparatus is enormously minimized. Then, too, the case can be made cylindrical and corrections for its attraction avoided. Until quartz fibres existed it would have been impossible to have made this reduction in the dimensions of the apparatus and retained the period of 5 to 10 minutes. The introduction of this invaluable new means of suspension is also due to Professor Boys. Another improvement in the form of the apparatus devised by him is the suspending of the small balls at different distances below the arm (the masses must be at corresponding levels), so that each mass acts practically on one ball only.

Boys showed to the Society a balance of this design; it had an arm of only 13 mm. in length, was 18.7 times as sensitive as that of Cavendish and behaved very satisfactorily. He proposed to prepare a balance of this kind especially suitable for absolute determinations and capable of determining the gravitation constant to 1 part in 10 000.

An account of his completed work (187 and 189) was read in 1894. For the details of this beautiful experiment and the ingenious way in which the apparatus was designed the original paper must be consulted. The general design was that of his earlier apparatus, but very great attention was given to the minutest details, and especially to the arrangements for measuring the dimensions. Some idea of the accuracy aimed at may be got from considering that in order to obtain a result correct to 1 in 10 000 it was necessary to measure the large masses to 1 in 10 000, the times to 1 in 20 000, some lengths to 1 in 20 000 and angles to 1 in 10 000. The dimensions finally used were, diameter of masses 2.25 and 4.25 in.; distance between masses *in plan* 4 and 6 in. ; distance between balls *in plan* 1 in. ; diameter of balls .2 and .25 in. ; difference of level between upper and lower balls 6 in. The masses were of lead formed under great pressure, and the balls of gold.

The moment of inertia of the beam was determined by finding the period when the balls were suspended from it, and when they were taken away and a cylindrical body of silver, equal in weight to the balls with their attachments, suspended from the middle of the beam. The apparatus was enclosed by a series of metallic screens to prevent temperature changes, and outside of all was a double-walled wooden box with the space between the walls filled with cotton-wool. The final result for the gravitation constant was 6.6576×10^{-8}, and for Δ, 5.5270. The last figure in each case has no significance, but Boys considered that the next to the last could not be more than 2 in error at the outside. He is still convinced that 1 part in 10 000 can be reached, but would increase the length of the beam to 5 cm., since the disturbing moments due to convection are proportional to the 5th power of the linear dimensions, not to the 7th as he had originally supposed. An excellent résumé of the experiment is to be found in the lecture delivered by Boys before the Royal Institution (188).

EOETVOES. A series of investigations upon gravitation is now under way by Prof. R. von Eötvös of Budapest. He has published a preliminary account (192) only of his experiments, but they promise to be very elaborate and exhaustive. His paper begins with a mathematical discussion of the space-variation of gravity as deduced from the potential function. He investigates the equipotential surface and the measurements necessary to determine the principal radii of curvature, the variation of gravity along the surface, and the variation perpendicular to the surface. The latter has already been measured with the pendulum, and by Jolly (144 and 145), Keller (152), Thiesen (179) and others with the common balance. For the measurement of the other quantities von Eötvös uses the torsion balance. This he makes in two forms: the first is of the same general type as that of Baily and is called the *Krümmungsvariometer*, since it is used to measure the difference of the reciprocals of the principal radii of curvature; the second is like that of Boys in that one ball is on one end of the rod and the other suspended 100 cm. below the other end by means of a wire, and is called the *Horizontalvariometer*. The peculiarity of these instruments is in the method devised for getting rid of convection currents; von Eötvös makes the case with double walls of thin metal with an air-space of from 5 to 10 mm. So steady is the motion that the balance can be used in any room in the laboratory, and even in the free air at night. The period is usually from 10 to 20 minutes, and the suspension wire is of platinum of 100 to 150 cm. length. The rod swings in a flat cylindrical box 40 cm. in diameter and 2 cm. deep.

Some investigations have been made of the variation of gravity in the neighbourhood of the hill Sághberg, where von Sterneck found great peculiarities. Some preliminary determinations have been made of the constant of gravitation also, with a result 6.65×10^{-8}. Von Eötvös speaks of the method employed as an entirely new one, but it is only a variation of that already employed by Reich, and later by Dr. Braun, the oscillation method. The instrument (the Krummungsvariometer) is set up between two pillars of lead, and the time of vibration observed both when the torsion rod is in the line joining the pillars and when it is perpendicular to this line. The paper is characterized by an almost total disregard of the work already done in the field of gravitation.

Eötvös gives a description of two new instruments for use in the study of gravitation. One he calls the *Gravitationcompensator ;* in design it is similar to the others, but the arm swings in a narrow tube. The tube is surrounded at each end by the compensating masses having the balls at their centres ; these masses are of the shape of a disc with two almost complete quadrants taken away, just enough being left to hold the remaining two quadrants together. By orienting these masses any amount of compensating attraction required can be produced. The other instrument is called the *Gravitationmultiplicator ;* underneath the torsion balance is a turn-table with the attracting mass ; when the ball has reached its maximum elongation in the direction of the mass, the latter is suddenly moved to the opposite side, and so on. From the difference of two successive elongations, and a knowledge of the damping, the amount of the attraction can be determined. This is rather similar to a piece of apparatus proposed by Joly (177), in 1890.

BRAUN. One of the latest and most elaborate determinations of the mean density of the earth is that made by Dr. Carl Braun, S.J., at Mariaschein in Bohemia (193). In its general form the apparatus is like that employed by Reich in his later experiments. Like Reich, too, he uses two distinct methods of finding his results, the deflection and the oscillation methods. The experiments of Dr. Braun differ however in several very important respects from those of Reich ; the dimensions of the apparatus are much reduced, the masses are suspended from wires and the deflection is determined differently ; but the respect in which it differs from all previous determinations is in the fact that the torsion rod swings in a partial vacuum of about 4 mm. of mercury. This was suggested by Faye (130) and by Boys, but no investigation of the kind had ever been made.

The experiments were begun in 1887. In a corner of a living-room a heavy stone slab was set into the stone wall; on this was a glass plate from which arose a brass tripod to carry the suspension wire, 1 m. long, and the torsion-rod; and on the plate fitted airtight a conical glass cover within which a vacuum could be made. The apparatus was so tight that the pressure inside did not change in 4 years. Suspended from a movable ring encircling the glass cover were the two masses, about 42 cm. apart; the masses used were two sets of spheres, one of brass weighing 5 kg. each, and the other of iron, 112 mm. in diameter, filled with mercury, and weighing about 9.15 kg. each. The torsion rod was a triangle of copper wires and the balls were suspended from its ends and lay in the same horizontal plane 24.6 cm. apart. Each ball was of gilded brass and weighed about 54 gr. In order to provide against temperature changes, the whole apparatus was surrounded by metal screens, cloth hangings and wooden enclosures.

The deflection method of observation is practically that of Cavendish, but the position of the centre was found rather differently. Dr. Braun observed the time of the passage in each direction across the wire of the telescope of several scale divisions near the centre, and took as the centre that point with reference to which the time of oscillation was the same in both directions. He found for the final corrected mean result $\Delta = 5.52962$.

In the oscillation method the period was determined when the masses were in the line joining the balls, and also when they were in a line at right angles to that direction. The final corrected mean result gave $\Delta = 5.52920$. The mean of all is $5.52945 \pm .0017$. The extremes were 5.5094 and 5.5511. The mean of all the results found in 1892 was 5.52770, and in 1894 was 5.53048.

The final result for the gravitation constant was 6.655213 $\times 10^{-8}$.

In an appendix is given a further discussion of the corrections to be made on account of damping. Dr. Braun found his former estimate to be in error, and after examination gave as the most probable final results, $\Delta = 5.52700 \pm .0014$, and $G = (6.65816 \pm .00168) \times 10^{-8}$.

A concise account of the work is given in *Nature* (197).

KOENIG, RICHARZ AND KRIGAR - MENZEL. In 1884, Professors A. Konig and F. Richarz proposed (159 and 160) to determine the gravitation constant by a method which is a modification of that used by Jolly (page 126). In the latter experiment the lower set of scale pans was 21 m. beneath the upper and differences of temperature were unavoidable. The improvement proposed was to have the sets of scale pans much closer together, to measure the change of weight with height after the manner of Jolly and then to insert between the upper and lower pans a huge block of lead with holes in it for the passage of the wires supporting the lower pans. A weighing was made with two nearly equal masses, one in the right upper, the other in the left lower pan ; then the former in the right lower is balanced against the latter in the left upper pan. From these weighings, taking account of the result of similar observations without the block, the value of 4 times the attraction of the block is determined, and from a comparison of this result with the calculated attraction, the gravitation constant can be determined. Professors König and Richarz seem to have hit upon the same idea independently of each other. In 1881, Keller proposed (167) a somewhat similar modification of the Jolly experiment. Professor A. M. Mayer suggested (161) the use of mercury instead of lead for the attracting mass, but König and Richarz replied (162) that Mayer had misunderstood the form of the experiment, and gave a lucid and simple explanation of their method.

In 1893, appeared a report (183 and 184) on the observations made to find the decrease of gravity with increase of height. A description is given of the balance and of the improvements introduced into it in order to overcome the liability to variation in its readings. The masses weighed against each other were 1 kg. each, and the balance had a sensitiveness of 1 part in 1 000 000. All exchange of weights was made automatically without opening the covers. The apparatus was carefully surrounded with metal screens to ward off temperature changes. It was set up in a bastion of the citadel at Spandau, and in consequence of the departure of König to accept a professorship at Berlin, Dr. Krigar-Menzel assisted in the carrying on of the research. The pans were 2.26 m. apart vertically. The change in gravity observed was .000006523 $\frac{m.}{sec.^2}$; whereas the calculated value was .00000697. The difference is ascribed to the local attraction of walls, etc.

The paper (198) embodying the final results was presented to the Berlin Academy in Dec., 1897. A most exhaustive examination had been made of the possible sources of error, and the devices for overcoming these difficulties were most ingenious and elaborate. In the cases where the sources of error could not be eliminated, as in the variations of temperature with time and place, the effect is carefully considered and allowed for. Observations were made continuously from Sept., 1890, to Feb., 1896, and from the elegance of the method and the time and care devoted to the working out of the result, this determination of the gravitation constant and the mean density of the earth must be taken as one of the very best.

The block of lead weighed 100 000 kg., was 200 cm. high and 210 cm. square and was built up out of bars of lead $10 \times 10 \times 30$ cm. on the top of a massive pier. The amount of the settling of the pier was measured and found to be not important, and the shape of the block was not distorted by its own pressure. The final value for G was $(6.685 \pm .011) \; 10^{-8}$, and for Δ, 5.505 $\pm .009$.

Professors Richarz and Krigar-Menzel published (191 and 199), in 1896, a condensed account of their work. Other abstracts are also to be found (185, pp. 64–5, 186$\frac{1}{2}$, pp. 26–7, and 194).

MINOR NOTICES. In 1889, Dr. W. Láska of Prague proposed (173) a method of finding the density of the earth. At the top of a rod projecting above a pendulum is a lens which is so close to a fixed plate of glass that Newton's rings are visible. A hollow ball near the bob of the pendulum is then filled with mercury and attracts the bob, bringing the lens nearer the plate ; an observation of the movement of the Newton's rings will measure the deflection of the bob. No further report has been published. An account of his method is given by Günther (196$\frac{1}{4}$, vol. 1, p. 197).

About the same time Professor Joly of Dublin suggested (177) a resonance method for the same purpose. A pendulum in a vacuous vessel has the same period as two massive ones kept going outside the vessel. The amplitude of the motion of the inner pendulum due to a given number of swings of the outer ones would give a measure of the constant of gravitation.

In 1895, Professor A. S. Mackenzie of Bryn Mawr College published an account (190) of some experiments with the Boys' form of torsion balance to determine whether the gravitational properties of crystals vary with direction. No such variation was found in the case of calc-spar, the crystal under investigation. He shewed further that the inverse square law holds good in the neighbourhood of a crystal to one-fifth per cent.

Two years later appeared an account (196) of an investigation by Professors Austin and Thwing of the University of Wisconsin to determine whether gravitational attraction is independent of the intervening medium, that is, whether there is a gravitational permeability. No effect was found due to the medium within the limits of error of the method.

At a meeting of the " Deutcher Naturforscher und Aerzte" in Brunswick, in 1897, Professor Drude read a paper (195) on action at a distance, which contains a very valuable account of the theory of gravitation, and should be consulted by any one wishing to find a brief résumé of that subject, and especially for a discussion of the velocity of propagation of gravitation.

The latest work on the laws of gravitation is that of Professors Poynting and Gray (200) on the search for a directive action of one quartz crystal on another. A small crystal was suspended and its time of rotative vibration noted ; a large crystal in the same horizontal plane was then rotated about a vertical axis through its centre with a period either equal to, or twice, that of the smaller crystal. If there were any directive action the small crystal should be set in vibration by forced oscillations ; no such effect was found.

TABLE OF RESULTS OF EXPERIMENTS

Date of experiment	Method	Experiment performed by	Result calculated by	Resulting value of Δ
1737-40	Mountain Pendulum	Bouguer	Author Saigey Author Sabine	4 7 ×density of Cordilleras. " 4.25 " " 4.39. Known to be in error. 4.77 Author's error having been corrected.
1821	" "	Carlini	Schmidt Giulio Saigey Knopf	4.837 4.95 6 15. 5 08.
1880	" "	Mendenhall	Author Fresdorf	5.77. 5.667.
1737-40	Plumb-line	Bouguer	Author Saigey	6 −7×density of Chimborazo. 4 62 " " " 1.83 " " " Using the best observations only.
1774-6	"	Maskelyne	Hutton Playfair	5. Approximately. 4.713.
1842	"	Puissant	Saigey	4.25.
1855	"	James and Clarke	Authors	5.316 ± 054.
1864	"	Pechmann	Author Author (corrected)	6.1311 ± 1557 6 352 ± 726. 5 448 ± 033. Cavendish gave 5.48, due to error in calculation
1797-8	Tors. pend.—Deflection	Cavendish	Schmidt Gosseln Babinet	5 52. 5 69 Treatment rather elementary. 5 5
1835-7	" "	Reich	Author Author Author	5.44 ±.0233. 5.4522. Using masses of iron 5.49 ± 020. Recalculated in 1852.
1838-41	" "	Baily	Author Saigey Cornu and Baille	5.6747 ±.0038. 5 52. Method dubious. 5 55. Author's method having been corrected

Year	Method	Experimenter		Result
1847–50	Tors. pend.—Deflection	Reich	Author	5.5832 ± 0149.
1870–8	" "	Cornu and Baille	Authors	5.56 in summer. ⎱ Early results. 5.50 in winter. ⎰ 5.56. Latest result.
1889–94	" "	Boys	Author	5.5270 ± .002 ....10⁸G = 6 6576 ±.002.
1887–96	" "	*Braun	Author	5.52962. ....10⁸G = 6.655.
1852	" " —Oscillation	Reich	Author	6.25.
1896	" "	Eötvös	Author	.............10⁸G = 6.65.
1887–96	" "	*Braun	Author	5 52920.........10⁸G = 6 665.
1826–8	Mine Pendulum	⎱ Airy, Whewell and Sheepshanks	Authors	6. Approximately.
1854	" "	Airy	⎱ Author ⎰ Haughton ⎰ Foie	6.566 ± .0182. 5.480. Treatment elementary. 6.439.
1882	" "	Sterneck	Author	⎱ 6.28. 5.01.
1883	" "	Sterneck	⎱ Author ⎰ Helmert	5.77. 5.71.
1885	" "	Sterneck	Author	5.66 ⎱ Increasing with depth of mine 6.66 ⎰ and with temperature. 7.15 7.60
1879–80	Common Balance	Jolly	Author	5.692 ±.068.
1878–90	" "	Poynting	Author	5 4934..........10⁸G = 6.6984.
1884–97	" "	⎱ König, Richarz and Krigar-Menzel	Authors	5.505 ± .009.....10⁸G = 6.685 ±.011.
1886–8	Pendulum Balance	Wilsing	Author	5.579 ± .012.
1893	Gravimeter	Berget	Author	5.41. ....10⁸G = 6.80.

*Braun's most probable results from all determinations combined. Δ = 5 52700 ± 0014
10⁸G = 6 65816 ± 00168.

144

THE LAWS OF GRAVITATION

BIBLIOGRAPHY

The numbers marked with an asterisk have not been consulted by the editor

1	1600	W. Gilbert.	De magnete magneticisque corporibus et de magno magnete tellure physiologia nova. London. 4ᵗᵒ.

1   1600   W. Gilbert.   De magnete magneticisque corporibus et de magno magnete tellure physiologia nova. London. 4ᵗᵒ.

2   1665   F. Bacon.   Opera omnia—Frankfurt ᵃ/ₘ. Folio.

3   1687   I. Newton.   Philosophiae naturalis principia mathematica. London. 4ᵗᵒ.

4   1705   R. Hooke.   Posthumous works. London. Folio.

5   1727   I. Newton.   De mundi systemate. London. 8ᵛᵒ.

6   1744   R Boyle.   Works, edited by Birch. 5 vol. London. Folio.

7   1749   P. Bouguer.   La figure de la terre. Paris. 4ᵗᵒ.

8   1751   Ch. M. de la Condamine.   Journal du voyage. Paris. 4ᵗᵒ.

8½*   1752–4   Ch. M. de la Condamine.   Supplément au journal historique. 2 parts. Paris. 4ᵗᵒ.

9   1754   P. Bouguer.   Sur la direction qu' affectent les fils-à-plomb. *Hist. de l'Acad Roy. des Sc. avec les Mém. de Math. et de Phys.*, 1–10 and 150–168.

10   1756   T. Birch.   The history of the Royal Society of London. 4 vol. London. 4ᵗᵒ.

11*   1769   J. Coultaud.   (Letter of 30 pp.). *Journ. des Sc. et des Beaux Arts.* June.

12*   1769   J. d'Alembert.   (Letter). *Journ. des Sc. et des Beaux Arts.* July.

13   1761–80   J. d'Alembert.   Opuscules mathématiques. 8 vol. Paris. 4ᵗᵒ.

14*   1771   J. P. David.   Dissertation sur la figure de la terre. La Haye. 8ᵛᵒ.

15*   1771   Mercier.   (Letter of 27 pp). *Journ. des Sc. et des Beaux Arts.* Dec.

16*   1772   G. L. Lesage.   Solutions des doutes. *Journ. des Sc. et des Beaux Arts.* April.

17*   1772   J. J. La Lande.   Remarques sur de nouvelles expériences de pesanteur. *Le Journal des Sçavans.* Aug.

18*   1772   de la P. de Roiffé.   Expérience du pendule de le Mercier aux Alpes du Valais. *Journal Encyclopédique*, **130**, 250.

19   1773   G. L. Lesage.   Lettre sur la fausseté de deux suites d'expériences. [*Rozier*] *Journ de Phys.*, **1**, 249–260.

20   1773   Father Bertier   (Letters). [*Rozier*] *Journ. de phys.*, **2**, 251–3 and 275.

21 1773 de la P. de Roiffé. Observations sur l'expérience de père Bertier. [*Rozier*] *Journ. de Phys.*, **2**, 374–8.

22 1773 G. L. Lesage. Réflexions sur une nouvelle expérience du Révérend Père Bertier. [*Rozier*] *Journ. de Phys.*, **2**, 378–81.

23 1774 J. P David and Fathers Cotte and Bertier. (Notice of their experiments). [*Rozier*] *Journ. de Phys.*, **4**, 338.

24 1774 J. P. David. Réponse aux réflexions de M. Lesage. [*Rozier*] *Journ. de Phys.*, **4**, 431–41.

25 1774 Abbé Rozier. Observations sur la lettre de Père Bertier. [*Rozier*] *Journ. de Phys.*, **4**, 454–61.

26* 1774 Father Bertier. (Account of experiments). *Journ. de Verdun*, 148–185.

27 1775 J. P. David. Sur la pesanteur des corps. [*Rozier*] *Journ. de Phys* , **5**, 129–139.

28 1775 Father Bertier. (Letter). [*Rozier*] *Journ. de Phys.*, **5**, 305–13.

29 1775 — (Account of exp'ts. made by com. of Acad. of Dijon). [*Rozier*] *Journ. de Phys.*, **5**, 314–26.

30 1775 Chev. de Dolomieu. Expériences sur la pesanteur des corps à différentes distances du centre de la terre. [*Rozier*] *Journ. de Phys.*, **6**, 1–5

31 1775 N. Maskelyne A proposal for measuring the attraction of some hill in this kingdom by astronomical observations. *Phil. Trans. Lond.*, 495–9.

32 1775 N. Maskelyne. An account of observations made on the mountain Schehallien for finding its attraction. *Phil. Trans. Lond* , 500–42.

33 1776 F. K. Achard. Bemerkungen über die von Herrn Bertier angestellten Versuche. *Beschäft. der Berl. Ges. Naturf. Freunde*, **2**, 1–11.

34 1776 G. L. Lesage. Expériences et vues sur l'intensité· de la pesanteur dans l'intérieur de la terre. [*Rozier*] *Journ. de Phys.*, **7**, 1–12.

35 1776 J. Pringle. Discours sur l'attraction des montagnes ; traduit par M. le Roy. [*Rozier*] *Journ. de Phys.*, **7**, 418–34.

36 1777 Father Bertier. Rétraction du Père Bertier de l'Oratoire, sur la conséquence qu'il a tiré de son expérience d'un corps, pesant plus dans un lieu haut que dans un bas. [*Rozier*] *Journ. de Phys.*, **9**, 460–6.

37 1779 C. Hutton. An account of the calculations made from the survey and measures taken at Schehallien in order to ascertain the mean density of the earth. *Phil. Trans. Lond.*, **68**, 689–788.

38 1780 C. Hutton. Calculations to determine at what point in the side of a hill its attraction will be the greatest. *Phil. Trans. Lond.*, 1–14.

39 1798 H. Cavendish. Experiments to determine the density of the earth. *Phil Trans. Lond.*, **88**, 469–526.

THE LAWS OF GRAVITATION

39½ 1799 — (Review of 39). *Bibl Brit.*, **11**, 233–41.

40 1799 L. W. Gilbert. Versuche, um die Dichtigkeit zu bestimmen, von Henry Cavendish, Esq. [*Gilbert*] *Ann. der Phys.*, **2**, 1–62.

41 1803 A. Motte. The mathematical principles of natural philosophy by Sir I. Newton, translated into English by Andrew Motte, to which are added Newton's system of the world, etc. W. Davis' edn. 3 vol. London. 8vo.

42 1806 H. W. Brandes. Theoretische Untersuchungen über die Oscillationen der Drehwaage bei Cavendish's Versuchen über die Attraction kleiner Massen. *Mag. fur den Neuesten Zustand der Naturkunde*, **12**, 300–310.

43 1809 F. X. von Zach. Ueber die Möglichkeit die Gestalt der Erde aus Gradmessungen zu bestimmen. *Monatl. Corresp.*, **20**, 3–9.

44 1810 F. X. von Zach. Ueber Densität der Erde und deren Einfluss auf geographische Ortsbestimmungen *Monatl. Corresp*, **21**, 293–310.

45 1811 C. Hutton. On the calculations for ascertaining the mean density of the earth. [*Tilloch*] *Phil. Mag.*, **38**, 112–6.

46 1811 J. Playfair. Account of a lithological survey of Schehallien. made in order to determine the specific gravity of the rocks which compose that mountain. *Phil. Trans. Lond.*, 347–77.

47 1812 C. Hutton. Tracts on mathematical and philosophical subjects. 3 vol. London. 8vo.

48 1813 L. W. Gilbert. Bericht von einer lithologischen Aufnahme des Schehallien, um das specifische Gewicht der Gebirgsarten desselben, und daraus die mittlere Dichtigkeit der Erde zu bestimmen, von J. Playfair, Esq. *Pogg Ann.*, **43**, 62–75.

49 1814 F. X. von Zach. L'attraction des montagnes, et ses effets sur les fils à plomb ou sur les niveaux des instruments d'astronomie. 2 vol. Avignon. 4to.

50 1815 N. M. Chompré. Expériences pour déterminer la densité de la terre ; par Henry Cavendish. Traduit de l'Anglais. *Journ. de l'Éc Roy. Polytechnique.* Cahier 17. **10**, 263–320.

51 1819 T. Young. Remarks on the probabilities of error in physical observations, and on the density of the earth. *Phil. Trans Lond.*, 70–95 ; *Misc Works*, **2**, 8–28.

52 1820 C. Hutton. (Letter to Laplace). [*Blainville*] *Journ. de Phys.*, **90**, 307–12.

53 1821 C. Hutton. On the mean density of the earth. [*Tilloch*] *Phil. Mag.*, **58**, 3–13.

54 1821 C. Hutton. (Same title as 53). *Phil Trans Lond.*, 276–292.

55 1824 F. Carlini. Osservazioni della lunghezza del pendolo semplice fatte all' altezza di mille tese sul livello del mare. *Eff Astr. di Milano,* app. 28–40.

56 1825 S. (Notice of 55). [*Férussac*] *Bull. des Sc. Math.,* **3**, 298–301.

57 1826 M. W. Drobisch. De vera lunae figura. Lipsiae. 12mo.

58 1827 E S(abine). An account of Prof. Carlini's experiments on Mont-Cenis. *Quart. Journ of Sc.,* **24**, 153–9.

59 1827 M. W. Drobisch. Ueber die in den Minen von Dolcoath in Cornwall neuerlich angestellten Pendelbeobachtungen. *Pogg. Ann.,* **10**, 444–456.

60 1827 — (Notice of Dolcoath expt). *Phil. Mag.,* [2], **1**, 385–6.

61 1825–45 J. S. T. Gehler. Physikalisches Wörterbuch. 22 vol. Leipzig 8vo.

62* 1828 — Account of experiments made at Dolcoath mine in Cornwall, in 1826 and 1828, for the purpose of determining the density of the earth. Cambridge. 8vo. Printed privately.

63 1828 M. W. Drobisch. Ausführlicher Bericht über mehrere in den Jahren 1826 und 1828 in den Minen von Dolcoath in Cornwall zur Bestimmung der mittleren Dichtigkeit der Erde angestellte Pendelversuche. *Pogg. Ann.,* **14**, 409–27.

64 1829–30 J C. E. Schmidt. Lehrbuch der mathematischen und physischen Geographie. 2 vol Göttingen. 8vo.

65 1833 S D. Poisson. Traité de mécanique 2^d ed^n 2 vol. Paris. 8vo.

66 1837 E de Beaumont. Extrait d un mémoire de M. Reich sur la densité de la terre. *Comp. Rend ,* **5**, 697–700.

67 1838 F. Reich. Versuche über die mittlere Dichtigkeit der Erde mittelst der Drehwage. Freiberg. 8vo.

68 1838 — On the repetition of the Cavendish experiment, for determining the mean density of the earth. *Phil. Mag.,* [3], **12**, 283–4.

69 1839 F Baily. (Same title as 68). *Mon. Not. Roy. Astr. Soc ,* **4**, 96–7.

70 1840 C. I. Giulio. Sur la détermination de la densité moyenne de la terre, déduite de l'observation du pendule faite à l'Hospice du Mont-Cenis par M. Carlini en Septembre, 1821. *Mem. Accad. Torino,* [2], **2**, 379–84

71 1840 L. F. Menabrea Calcul de la densité de la terre. *Mem. Accad. Torino,* [2], **2**, 305–68

72 1840 A. G. Calcul de la densité de la terre, par L. F. Menabrea *Bibl. Univ. de Genève,* [*nouv.*], **27**, 168–75.

73 1841 L. F. Menabrea. On Cavendish's experiment *Phil Mag.,* [3], **19**, 62–3.

74 1842 J. F. Saigey. Densité du globe. *Rev Scient. et Ind.,* [*Quesneville*], **11**, 149–60 and 242–53, and **12**, 373–88.

75 1842 F. Baily. An account of some experiments with the torsion-rod, for determining the mean density of the earth. *Phil. Mag* , [3]. **21**, 111–21.

76 1842 F. Baily. Résultats de quelques expériences faites avec la balance de torsion, pour déterminer la densité moyenne de la terre. *Ann. de Chim. et de Phys.*, [3], **5**, 338–53.

77 1842 F. Baily. Bericht von einigen Versuchen mit der Drehwage zur Bestimmung der mittleren Dichtigkeit der Erde. *Pogg. Ann.*, **57**, 453–67.

78 1843 F. Baily. (Same title as 75). *Mon. Not Roy. Astr. Soc.*, **5**, 188 and 197–206.

79 1843 F. Baily. Experiments with the torsion-rod for determining the mean density of the earth *Mem Roy. Astr. Soc* , **14**, 1–120 and i –ccxlviii.

80 1843 A. G (Same title as 76). *Bibl. Univ. de Genève*, [*nouv*], **43**, 177–81.

80½ 1845 C. A F. Peters. Von den kleinen Ablenkungen der Lothlinie und des Niveaus, welche durch die Anziehungen der Sonne, des Mondes, und einiger terrestrischen Gegenstände hervorgebracht werden *Astr. Nach.*, **22**, 33–42.

81 1847 G. W. Hearn. On the cause of the discrepancies observed by Mr Baily with the Cavendish apparatus for determining the mean density of the earth. *Phil. Trans. Lond* , 217–29

82 1849 E. Sabine. Cosmos, by Alexander von Humboldt, translated under the superintendence of Lieut.-Col. Edward Sabine. 6th Edn. **1**. London. 8vo.

83 1852 F. Reich. Neue Versuche mit der Drehwaage. *Leip. Abh. math. phy. cl.*, **1**, 383–430.

84 1852 — Neue Versuche über die mittlere Dichtigkeit der Erde, von F. Reich. *Pogg. Ann.*, **85**, 189–98.

85 1852 Schaar. Rapport de M. Schaar sur un mémoire de M. Montigny relatif aux expériences pour déterminer la densité de la terre. *Bull Acad. Roy. Belg* , **19**, pt. 2, 476–81.

86 1853 — Nouvelles expériences sur la densité moyenne de la terre. *Ann. de Chim. et de Phys.*, [3], **38**, 382–3.

87 1853 F. Reich. New experiments on the mean density of the earth. *Phil Mag* , [4], **5**, 154–9.

88 1855 G. B. Airy. (Report on Harton expts.). *Mon. Not. Roy.*, *Astr. Soc.*, **15**, 35–6.

89 1855 G. B. Airy. Note respecting the recent experiments in the Harton Colliery. *Mon. Not. Roy. Astr. Soc.*, **15**, 46.

90 1855 — (Report on Harton expts.). *Mon. Not. Roy. Astr. Soc.*, **15**, 125–6.

91 1855 — Note sur les observations du pendule exécutées dans les mines de Harton pour déterminer la densité moyenne de la terre ; par M. Airy. *Ann. de Chim. et de Phys.*, [3], **43**, 381-3.

92 1855 — Extrait du rapport présenté à la 35me séance anniversaire de la Société Royale Astronomique de Londres par le conseil de cette société le 9 Fevrier, 1855. *Arch. des Sc. Phys. et Nat.*, **29**, 188-191.

93 1855 J. H. Pratt. On the attraction of the Himalaya mountains, and of the elevated regions beyond them, upon the plumb-line in India. *Phil. Trans. Lond.*, **145**, 53-100.

94 1855 G. B. Airy On the computation of the effect of the attraction of mountain-masses, as disturbing the apparent astronomical latitude of stations in geodetic surveys. *Phil. Trans. Lond.*, **145**, 101-4.

95 1855 T. Young. Miscellaneous works and life, by Peacock and Leitch. 4 vol. London. 8vo.

96 1856 J. H. Pratt. (Same title as 93). *Mon. Not. Roy. Astr. Soc.*, **16**, 36-41 and 104-5.

97 1856 G. B. Airy. (Same title as 94). *Mon. Not Roy. Astr. Soc.*, **16**, 42-43.

98 1856 — (Report on Harton expts.) *Mon. Not. Roy. Astr. Soc.*, **16**, 104.

99 1856 J. H Pratt. On the effect of local attraction upon the plumb-line at stations on the English arc of the meridian, between Dunnose and Burleigh Moor ; and a method of computing its amount. *Phil Trans. Lond*, **146**, 31-52.

100 1856 G. B. Airy. Account of pendulum experiments undertaken in the Harton Colliery, for the purpose of determining the mean density of the earth. *Phil. Trans. Lond.*, **146**, 297-342.

101 1856 G. B. Airy. Supplement to the "account of pendulum experiments undertaken in the Harton Colliery " ; being an account of experiments undertaken to determine the correction for the temperature of the pendulum. *Phil. Trans. Lond.*, **146**, 343-55.

102 1856 G. G. Stokes. (Addendum to 101 ; on the effect of the earth's rotation and ellipticity in modifying the numerical results of the Harton experiment). *Phil. Trans. Lond.*, **146**, 353-5.

103 1856 H. James and A. R. Clarke. On the deflection of the plumb-line at Arthur's Seat, and the mean specific gravity of the earth. *Phil. Trans. Lond.*, **146**, 591-606.

104	1856	H. James.	On the figure, dimensions and mean specific gravity of the earth, as derived from the ordnance trigonometrical survey of Great Britain and Ireland. *Phil. Trans. Lond.*, **146**, 607–26.
105	1856	—	Ueber die in der Kohlengrube von Harton zur Bestimmung der mittleren Dichte der Erde unternommenen Pendelbeobachtungen ; von G. B. Airy. *Pogg Ann* , **97**, 599–605.
106	1856	S. Haughton.	On the density of the earth, deduced from the experiments of the Astronomer Royal, in the Harton coal-pit. *Phil. Mag.*, [4], **12**, 50–1.
107	1856	G. B. Airy.	(Same title as 100). *Phil. Mag.*, [4], **12**, 228–31.
108	1856	H. James.	Account of the observations and computations made for the purpose of ascertaining the amount of the deflection of the plumb-line at Arthur's Seat, and the mean specific gravity of the earth. *Phil. Mag.*, [4], **12**, 314–6.
109	1856	G. B. Airy.	(Same title as 101). *Phil. Mag.*, [4], **12**, 467–8.
110	1856	—	Ueber die Dichtigkeit der Erde, hergeleitet aus den Versuchen des Königl. Astronomen (Hrn. Airy) in der Kohlengrube Harton ; von Sr. Ehrwürd. Samuel Haughton, Fellow des Trinity College in Dublin. *Pogg. Ann* , **99**, 332–4.
111	1856	G. B. Airy.	On the pendulum experiments lately made in the Harton Colliery, for ascertaining the mean density of the earth. *Am. Journ. Sc.*, [2], **21**, 359–64.
112	1857	E. R.	Mémoire sur les expériences enterprises dans la mine de Harton pour déterminer la densité moyenne de la terre, par G. B. Airy. *Arch. des Sc. Phys. et Nat.*, **35**, 15–29.
113	1857	—	Ueber die Dichtigkeit der Erde, hergeleitet aus den Pendelbeobachtungen des Herrn Airy in der Kohlengrube Harton von Herrn S Haughton, Fellow am Trinity-College in Dublin. *Zeit. für Math. u. Phys.*, **2**, 68–70.
114	1857	—	Ueber die Bestimmung der mittleren Dichtigkeit der Erde. *Zeit für Math. u. Phys.*. **2**, 128–30.
115	1857	—	(Same title as 103) *Proc. Roy. Soc. Edin.*, **3**, 364–6.
116	1857	—	(Notice of 106). *Am. Journ. Sc.*, [2], **24**, 158.
117	1857	H. James.	(Same title as 104). *Phil Mag.*, [4], **13**, 129–32.
118	1857	W. S. Jacob.	On the causes of the great variation among the different measures of the earth's mean density. *Phil. Mag.*, [4], **13**, 525–8.
119	1857	G. B. Airy.	(Same title as 101). *Proc. Roy. Soc. Lond.*, **8**, 58–9.
120	1857	H. James.	(Same title as 104). *Proc. Roy. Soc. Lond.*, **8**, 111–6.

121 1857 W. S. Jacob. (Same title as 118). *Proc. Roy. Soc. Lond.*, **8**, 295–9.

122 1858 G. B. Airy. (Same title as 111) *Proc. Roy. Inst.*, **2**, 17–22.

123 1858 — (Same title as 103). *Mon. Not. Roy. Astr. Soc.*, **18**, 220.

124 1858 — (Same title as 104). *Mon. Not. Roy Astr. Soc.*, **18**, 220–2.

125 1858 H. James and A R Clarke. Ordnance trigonometrical Survey of Great Britain and Ireland. Account of the observations and calculations of the principal triangulation ; and of the figure, dimensions and mean specific gravity of the earth as derived therefrom. 2 vol. London. 4to.

126 1859 — (Same title as 125). *Mon. Not. Roy. Astr. Soc.*, **19**, 194–9.

127 1859 P. F. J. Gosselin. Nouvel examen sur la densité moyenne de la terre. *Mem. Acad. Imp. de Metz*, [2], **7**, 469–85.

128* 1859–60 E. Sergent. Sulla densitá della materia nell intorno del globo, e sulla potenza della crosta terrestre. *Atti della Soc. Ital. di Sc. Nat. Milano*, **2**, 169–175.

129 1861 O. Struve. Ueber einen von General Schubert an die Akademie gerichteten Antrag betreffend die Russisch Scandinavische Meridian - Gradmessung. *Bull. Acad. St. Petersb. phys. math. cl.*, **3**, 395–424.

130 1863 H. A. E. A. Faye. Sur les instruments géodésiques et sur la densité moyenne de la terre. *Comp. Rend.*, **56**, 557–66.

131 1864 E. Pechmann. Die Abweichung der Lothlinie bei astronomischen Beobachtungsstationen und ihre Berechnung als Erforderniss einer Gradmessung. *Denkschr. Acad. Wiss. Wien. math.-naturw. cl.*, **22**, 41–88.

132 1864 J. Babinet. Note sur le calcul de l'expérience de Cavendish, relative à la masse et à la densité moyenne de la terre. *Cosmos*, **24**, 543–5.

133 1865 J H. Pratt. A treatise on attractions, Laplace's functions, and the figure of the earth. 3^d. Ed^n. Cambridge and London. 8vo.

134 1865 H. Scheffler. Ueber die mittlere Dichtigkeit der Erde. *Zeit. für Math. u. Phys.*, **10**, 224–7.

135 1869 A. Schell. Ueber die Bestimmung der mittleren Dichtigkeit der Erde. Göttingen. 4to.

136 1872 F. Folie. Sur le calcul de la densité moyenne de la terre, d'après les observations d'Airy. *Bull. Acad. Roy. Belg.*, [2], **33**, 369–372 and 389–409.

137 1873 A. Cornu et J. B. Baille. Détermination nouvelle de la constante de l'attraction et de la densité moyenne de la terre. *Comp. Rend*, **76**, 954–8.

THE LAWS OF GRAVITATION

138* 1873 — (Notice of 137). *Bull. Hebd. de l'Assoc. Scient. de France*, [1], **12**, 70.

139   1873   A. Cornu and J. B. Baille. Mutual determination of the constant of attraction, and of the mean density of the earth. *Chemical News*, **27**, 211.

140   1873   I. Todhunter. A history of the mathematical theories of attraction and the figure of the earth, from the time of Newton to that of Laplace. 2 vol. London. 8vo.

141   1878   A. Cornu et J. B. Baille. Étude de la résistance de l'air dans la balance de torsion. *Comp. Rend.*, **86**, 571–4.

142   1878   A. Cornu et J. B. Baille. Sur la mesure de la densité moyenne de la terre *Comp. Rend.*, **86**, 699–702.

143   1878   A Cornu et J. B. Baille. Influence des termes proportionels au carré des écarts, dans le mouvement oscillatoire de la balance de torsion. *Comp. Rend.*, **86**, 1001–4.

144   1878   Ph. von Jolly. Die Anwendung der Waage auf Probleme der Gravitation Part 1. *Abh. Bay. Akad. Wiss. cl.* 2, **13**, *Abth.* 1, 157–176.

145   1878   Ph. von Jolly. (Same title as 144). *Wied. Ann.*, **5**, 112–34.

146   1879   J. H. Poynting. On a method of using the balance with great delicacy, and on its employment to determine the mean density of the earth. *Proc. Roy. Soc. Lond.*, **28**, 2–35.

146½ 1880   H. A. E. A. Faye. Sur les variations séculaires de la figure mathématique de la terre. *Comp. Rend.*, **90**, 1185–91.

147   1880   H. A. E. A. Faye. Sur la réduction des observations du pendule au niveau de la mer. *Comp. Rend.*, **90**, 1443–6.

148   1880–4 F. R. Helmert. Die mathematischen und physikalischen Theorieen der höheren Geodäsie. 2 vol. Leipzig. 8vo.

148½ 1880–5 O. Zanotti-Bianco. Il problema meccanico della figura della terra. 2 parts. Firenze-Torino-Roma. 8vo.

149   1880   A. R. Clarke. Geodesy. Oxford. 8vo.

149½* 1880   O. Knopf. Ueber die Methoden zur Bestimmung der mittleren Dichtigkeit der Erde. Jena.

150   1880   T. C. Mendenhall. Determination of the acceleration due to the force of gravity, at Tokio, Japan. *Am. Journ. Sc.*, [3], **20**, 124–32.

151   1881   T. C. Mendenhall. On a determination of the force of gravity at the summit of Fujiyama, Japan. *Am. Journ. Sc.*, [3], **21**, 99–103.

152   1881   F. Keller. Sulla diminuzione della gravitá coll'altezza. *Atti Accad. Lincei. Mem. cl. sc.*, [3], **9**, 103–17.

153   1881   Ph. von Jolly. (Same title as 144). Part 2. *Abh. Bay. Akad. Wiss cl.* 2, **14**, *Abth.* 2, 3–26.

MEMOIRS ON

154   1881   Ph. von Jolly. (Same title as 153).   *Wied. Ann.*, **14**, 331–55.

154½ 1882   J. G. Wallentin.   Ueber die Methoden zur Bestimmung der mittleren Dichte der Erde und eine neue diesbezügliche Anwendung der Wage. *Humboldt*, **1**, 212–7.

155   1882   R. von Sterneck.   Untersuchungen über die Schwere im Innern der Erde. *Mitth. Mil. Geog. Inst. Wien*, **2**, 77–120.

156   1883   R. von Sterneck.   Wiederholung der Untersuchungen über die Schwere im Innern der Erde. *Mitth. Mil.-Geog. Inst. Wien*, **3**, 59–94.

157   1883   J. B. Baille.   Sur la résistance de l'air dans les mouvements oscillatoires très lents. *Comp. Rend.*, **96**, 1493–5.

158   1884   R. von Sterneck.   Untersuchungen über die Schwere auf der Erde. *Mitth. Mil.-Geog. Inst. Wien*, **4**, 89–155.

159   1884   A. König and F. Richarz.   Eine neue Methode zur Bestimmung der Gravitationsconstante. *Sitzungsb. Akad. Wiss. Berlin*, 1203–5.

160   1885   A. König and F. Richarz. (Same title as 159).   *Wied. Ann.*, **24**, 664–8.

161   1885   A. M. Mayer.   Methods of determining the density of the earth. *Nature*, **31**, 408–9.

162   1885   A. König and F. Richarz.   Remarks on our method of determining the mean density of the earth. *Nature*, **31**, 484.

163   1885   J. Wilsing.   Ueber die Anwendung des Pendels zur Bestimmung der mittleren Dichtigkeit der Erde. *Sitzungsb. Akad. Wiss. Berlin, Hbbd.* 1, 13–15.

164   1885   R. von Sterneck.   Fortsetzung der Untersuchungen über die Schwere auf der Erde. *Mitth. Mil.-Geog. Inst. Wien*, **5**, 77–105.

165   1886   R. von Sterneck. (Same title as 155).   *Mitth. Mil.-Geog. Inst. Wien*, **6**, 97–119.

166   1886   W. M. Hicks.   On some irregularities in the values of the mean density of the earth, as determined by Baily. *Proc. Cam. Phil. Soc.*, **5**, pt. 2, 156–61.

167   1886   F. Keller.   Sul metodo di Jolly per la determinazione della densitá media della terra. *Atti Accad. Lincei. Rend.*, [4], **2**, 145–9.

168   1887   F. Keller.   Sulla deviazione del filo a piombo prodotta dal prosciugamento del Lago di Fucino. *Atti Accad. Lincei. Rend.*, [4], **3**, 493–501.

169   1887   J. Wilsing.   Mittheilung über die Resultate von Pendelbeobachtungen zur Bestimmung der mittleren Dichtigkeit der Erde. *Sitzungsb. Akad. Wiss. Berlin, Hbbd.* 1, 327–34.

170   1887   J. Wilsing.   Bestimmung der mittleren Dichtigkeit der Erde mit Hülfe eines Pendelapparates. *Publ. Astrophys. Obs. Potsdam*, **6**, Stück 2, 35–127.

THE LAWS OF GRAVITATION

171 1888 J. H. Gore. Determination of the mean density of the earth by means of a pendulum principle, by J. Wilsing, translated and condensed. Smithsonian Rep. 1888. 635–46.

172 1889 J. Wilsing. (Same title as 170). *Publ. Astrophys. Obs. Potsdam*, **6**, Stück 3, 133–91.

173 1889 W. Láska. Ueber einen neuen Apparat zur Bestimmung der Erddichte. *Zeit. für Inst. - Kunde.*, **9**, 354–5.

174 1889 J. H. Gore. A bibliography of geodesy. Washington. 4to. App. to U. S. Coast and Geod. Surv. Rep. for 1887.

175 1889 C. V. Boys. On the Cavendish Experiment. *Proc. Roy. Soc. Lond.*, **46**, 253–68.

176 1889–90 C. V. Boys. (Same title as 175). *Nature*, **41**, 155–9.

177 1889–90 J. Joly. (Report of meeting of Univ. Exptl. Assoc. Dublin). *Nature*, **41**, 256.

178 1889–91 — Collection de mémoires relatifs à la physique, publiés par la Société Française de Physique, **4** and **5**. Paris. 8vo.

179 1890 Thiesen. Détermination de la variation de la pesanteur avec la hauteur. *Trav. et Mem. du Bur. Internat. des Poids et Mes.*, **7**, 3–32.

180 1891 J. H. Poynting. On a determination of the mean density of the earth and the gravitation constant by means of the common balance. *Phil. Trans. Lond.*, [A], **182**, 565–656.

181 1893 A. Berget. Détermination expérimentale de la constante de l'attraction universelle, ainsi que de la masse et de la densité de la terre. *Comp. Rend.*, **116**, 1501–3.

182 1893 Gouy. Sur la réalisation des temperatures constantes. *Comp. Rend.*, **117**, 96–7.

183 1893 F. Richarz und O. Krigar-Menzel. Die Abnahme der Schwere mit der Höhe bestimmt durch Wägungen. *Sitzungsb. Akad. Wiss. Berlin*, 163–83.

184 1894 F. Richarz und O. Krigar-Menzel. (Same title as 183). *Wied. Ann.*, **51**, 559–83.

185 1894 J. H. Poynting. The mean density of the earth. London. 8vo.

186 1894 J. H. Poynting. A history of the methods of weighing the earth. *Proc. Birmingham Nat. Hist. and Phil. Soc.*, **9**, 1–23.

186½ 1894 G. Fresdorf. Die Methoden zur Bestimmung der mittleren Dichte der Erde. *Wiss. Beilage zum Jahresb. des Gym. zu Weissenburg i. Elsass*.

187 1894 C. V. Boys. On the Newtonian constant of gravitation. *Proc. Roy. Soc. Lond.*, **56**, 131–2.

188 1894 C. V. Boys. (Same title as 187). *Nature*, **50**, 330–4, 366–8, 417–9 and 571.

189　1895　C. V. Boys.　(Same title as 187).　*Phil. Trans. Lond.*, [A], **186**, 1–72.

190　1895　A. S. Mackenzie.　On the attractions of crystalline and isotropic masses at small distances. *Phy. Rev.*, **2**, 321–43.

191　1896　F. Richarz und O. Krigar-Menzel.　Gravitationsconstante und mittlere Dichtigkeit der Erde, bestimmt durch Wägungen. *Sitzungsb. Akad. Wiss. Berlin*, 1305–18.

192　1896　R. von Eötvös.　Untersuchungen über Gravitation und Erdmagnetismus. *Wied. Ann.*, **59**, 354–400.

193　1896　C. Braun.　Die Gravitationsconstante, die Masse und mittlere Dichte der Erde nach einer neuen experimentellen Bestimmung. *Denkschr. Akad. Wiss. Wien. math.-naturw. cl.*, **64**, 187–258c.

194　1896–7　—　The gravitation constant and the mean density of the earth. *Nature*, **55**, 296.

195　1897　P. Drude.　Ueber Fernewirkungen. *Wied. Ann.*, **62**, i.–xlix.

196　1897　L. W. Austin and C. B. Thwing.　An experimental research on gravitational permeability. *Phy. Rev.*, **5**, 294–300.

196½　1897–9　S. Günther. Handbuch der Geophysik. 2 vol. Stuttgart. 8vo.

197　1897　J. H. P(oynting).　A new determination of the gravitation constant and the mean density of the earth. *Nature*, **56**, 127–8.

198　1898　F. Richarz und O. Krigar-Menzel.　Bestimmung der Gravitationsconstante und mittleren Dichtigkeit der Erde durch Wägungen. *Anhang Abh. Akad. Wiss. Berlin*, 1–196.

199　1898　F. Richarz und O. Krigar-Menzel.　(Same title as 191). *Wied. Ann.*, **66**, 177–193.

200　1899　J. H. Poynting and P. L. Gray.　An experiment in search of a directive action of one quartz crystal on another. *Phil. Trans. Lond.*, [A], **192**, 245–56.

INDEX

A

Achard, 49.

Airy, 5, 106, 113, 118, 119, 121–124, 128–130; Theory of Cavendish Experiment, 106, 118; Dolcoath Experiments, 113; Harton Experiments, 121.

Arthur's Seat, 118, 123, 124.

Attraction, Newton's Theorems on, 9; Newton's Error in Calculation of, 16, 17; Primitive, 27; Of a Plateau, 29–32; Of a Spherical Segment, Calculated by Newton, 17; by Carlini, Schmidt and Giulio, 111, 112; Shown by Deflection of Plumb-line, 33–43; Of Chimborazo, 34, 39; Of Schehallien, 43, 53–56; Due to Tides, 44, 134; Of any Hill, Calculated by Hutton, 54; Of the Great Pyramid, 55; Local, 56, 122–124, 126, 134, 135, 141; Of Mount Mimet, 56; Of Mass Beneath Earth's Surface, 56, 122, 123; Of Arthur's Seat, 118, 123, 124; Of Évaux, 118; Of a Cone, 128; Of an Infinite Plane, 135.

Austin and Thwing, 142.

B

Babinet, 106.

Bacon, 1, 2, 5, 49, 113.

Baily, 100, 105, 106, 115–120, 125, 131–133, 137; Cavendish Experiment Criticized by, 105; Error of, Pointed out by Cornu and Baille, 119; Anomalies in Results of, and their Explanations, 118, 119.

Balance, Experiments with Beam, 2–5, 48, 49, 125, 132, 140; Experiments with Torsion, 59–105, 114–121, 124, 135, 137–139, 142; Michell Devised Torsion, 60; Experiments with Pendulum, 131, 132.

Bauernfeind, 124.

Beaumont, 116.

Berget, 124, 134, 135.

Bertier, 47–49.

Boscovitch, 134.

Bouguer, 5, 21, 23–25, 27, 32, 33, 36, 39–44, 47, 53, 56, 130, 134; On Tides, 44, 134; Life of, 44; First to Take Account of Buoyancy of Air, 26.

Boyle, 4.

Boys, 106, 135–137, 139, 142.

Brandes, 105, 106, 120; Theory of Cavendish Experiment, 106; Theory of Oscillation Method, 105, 106, 120.

Braun, 106, 138, 139.

C

Carlini, 111–113, 128.

Cavendish, 54, 55, 59, 90, 91, 98, 100, 105–107, 114–116, 118, 119, 125, 135, 136, 139; Error in Calculation of, 100, 105; Life of, 107.

Chimborazo, 22, 34, 39–41, 43.

Clarke, 56, 118, 123, 124.

Condamine, de la, 21, 28, 32, 36, 39–41, 43, 44; Pendulum Experiments of, 28; Method of, for Doubling Deflection of Plumb-line, 36.

Cornu and Baille, 66, 106, 115, 119, 124, 125, 131, 135.

Cotte, 48.

Cotton, 4.

Coulomb Balance, First Proposed by Michell, 60.

Coultaud, 47, 48, 111.

D

D'Alembert, 31, 47.

Damping, Method of Finding Δ, 138, 141, 142; Effect of, 139.

157

INDEX

David, 47–49.

Deflection, of Arm of Torsion Balance, How Measured, by Cavendish, 64, 98 ; by Reich, 116, 119 ; by Baily, 117, 119, 132, 133 ; by Braun, 138 ; Affects the Period, 97 ; Error in Baily's Method of Observing, 119, 125 ; Multiplied by Poynting, 132, 133.

Descartes, 2, 49 ; Suggested Method of Measuring Gravity, 2.

Dimensions of Torsion Balance, Effects of, 125, 135, 137, 138.

Dolcoath, 113, 121.

Dolomieu, 49.

Drobisch, 113, 114.

Drude, 142.

E

Eötvös, 106, 137, 138.

F

Faye, 31, 124, 130, 139 ; Compensation Theory of, Correction of " Dr. Young's Rule," 31, 124.

Ferrel, 130.

Flotation Theory, 31, 124.

Folie, 123.

Forbes, 117, 118, 120.

Forced Vibrations, 138, 141, 142.

Fresdorf, 56, 112, 116, 123, 127, 128, 131, 132, 135.

Fujiyama, 127, 128.

G

Gilbert, Dr., 1, 5, 49.

Gilbert, L. W., 105.

Giulio, 112.

Gore, 124, 132.

Gosselin, 106.

Gouy, 135.

Gravimeter, 135.

Gravitation, Early Conceptions of, 1, 49, 56 ; Early Experiments on, by Members of Royal Society, 2–5 ; As Explanation of Planetary Motion, by Newton, 2, 10–19 ; Magnetic Theory of, 1, 4, 5, 12 ; Hooke's Ideas Concerning, 5, 6 ; Compensator, 138 ; Multiplicator, 138 ; Permeability, 142 ; Velocity of Propagation of, 142.

Gravity, Proposed Experiment on, by Bacon, 1 ; by Descartes, 2 ;

Decrease of, with Height, 27–33, 47–49, 111–113, 126–128, 130, 137, 140 ; Law of Increase of, with Depth, 129, 130 ; Increase of, with Temperature, 131 ; Mathematical Discussion of, from Potential, 137.

Gray, 142.

Günther, 131, 141.

H

Harton Colliery, 5, 121, 122

Haughton, 122.

Hearn, 118, 120.

Helmert, 54, 56, 124, 127, 129.

Hicks, 119, 131.

Hooke, 2, 4, 5.

Horizontalvariometer, 137.

Humboldt, 56.

Hutton, 54, 55, 90, 100, 105, 106, 118, 124.

J

Jacob, 56, 123.

James and Clarke, 56, 118, 123, 124.

Jolly, 125, 126, 137, 140.

Joly, 138, 141.

K

Keller, 124, 127, 135, 137, 140.

Kepler, 1, 2, 49.

Knopf, 113, 122.

König, 140.

Krigar-Menzel, 140, 141.

Krümmungsvariometer, 137, 138.

L

Lalande, 48.

Láska, 141.

Law, of the Distance, 2, 9, 29, 47, 101, 126, 142 ; Of the Masses, 13, 32 ; Of the Material, 12, 142 ; Of the Medium, 142.

Lesage, 2, 48, 49.

M

Mackenzie, 142.

Magnetism, Gilbert's Explanation of Gravitation by, 1, 4, 5 ; Contrasted with Gravitation, 12 ; Tests for Effects of, by Cavendish, 67, 68, 75, 76 ; by Reich, 116, 120 ; Suggested by Hearn to Account for Anomalies in Baily's Results, 118–120.

INDEX

Maskelyne, 17, 43, 53–56, 101, 106.
Mayer, 140.
Menabrea, 55, 66, 106.
Mendenhall, 124, 127, 128.
Mercier, 47, 48, 111.
Michell, 59, 60, 61.
Mine Experiments, 1, 2, 4, 5, 49, 113, 121, 128, 131.
Montigny, 119.
Muncke, 55, 105, 106.

N

Newton, 2, 6, 7, 9, 14–17, 19, 39, 43, 47–49, 56, 107, 110, 124, 126, 141; Explanation of Planetary Motions by, 2, 10; Pendulum Experiments of, 11, 15; Guess as to Value of Δ by, 14; Errors in Calculations of, 16, 17; Indicates Methods of Finding Δ, 17; Calculates Attraction of a Mountain, 17; Life of, 19; Attempts to Upset Theory of, 47.

P

Pechmann, 124.
Pendulum, Experiment with, Proposed by Bacon, 1; by Hooke, 5; Experiments with, by Newton, 10, 11, 15; by Bouguer, 24–33; by Coultaud and Mercier, 47; by Carlini, 111; by Airy, 113, 121; by Mendenhall, 127; by Sterneck, 128–131; by Láska, 141; Correction for Buoyancy of Air on, First Used, 26; Correction for, Due to Resistance of Air, 27, 66; Methods of Comparing One with Another, 113, 121, 128–130; Balance, 131.
Peters, 55, 124.
Playfair, 55, 102.
Plumb-line, Deflection of, Observed at Chimborazo, 33–43; at Schehallien, 43, 53–56; at Arthur's Seat, 123; at Évaux, 118, 124; in Tyrol, 124; Deflection of, Calculated for Chimborazo, 34; how to Observe, 35–39; 53; by Tides, 44, 134, 135.
Poisson, 31, 66
Power, 2–5, 49.
Poynting, 16, 36, 44, 106, 112, 116, 118, 119, 123–125, 127, 128, 131–134, 142.
Pratt, 124.
Pringle, 56.
Puissant, 118.
Pyramid, Attraction of the Great, 55.

R

Reich, 91, 106, 114–121, 125, 138.
Resistance of Air, Discussed by Bouguer, 27; by Cavendish, 65–67; by Poisson, Menabrea, and Cornu and Baille, 66, 106, 125.
Richarz, 140, 141.
Robison, 134.
Roiffé, 48.
Royal Society, Experiments by Members of, 2–6, 48.
Rozier, 49.

S

Sabine, 112.
Saigey, 32, 43, 54, 112, 113, 118, 124; Correction of Peruvian Pendulum Experiments by, 32, 43.
Schaar, 119.
Scheffler, 123.
Schehallien, 43, 53–55, 101, 112, 118, 123.
Schell, 56, 112, 116, 118, 123.
Schmidt, 44, 55, 106, 112.
Schubert, 124.
Sheepshanks, 113.
St. Paul's Cathedral, Experiments at, 4, 5.
Sterneck, 128–131, 137.
Stokes, 122.
Struve, 124, 134.

T

Temperature, Effects of, on Torsion Balance, Discussed by Cavendish, 60, 76–80; by Reich, 114; by Baily, 116, 117; by Hicks and Poynting, 119; by Boys, 135, 136; by Eötvös, 137; by Braun, 139; Change of Δ with, 125, 131; Limit of Constancy of, 135.
Thiesen, 137.
Thomson and Tait, 134.
Tides, Action of, on Plumb-line, 44, 134.
Time of Vibration, How Found by Cavendish, 64–67, 70; by Reich, 115, 119; by Baily, 117; by Mendenhall, 127; As Affected by Deflection, 97; As Affected by Convection Currents, 80, 100, 134, 137; Δ Found From, 105, 106, 120, 138, 139, 141.
Todhunter, 16, 44, 56, 66.

159

INDEX

U

Ulloa, 25, 39, 40.

V

Vacuum, Experiment Made in, 138, 141.

W

Wallentin, 127, 134.
Westminster Abbey, **Experiments** Made at, 2, 3, 5.
Whewell, 113.

Wilsing, 131, 132.

Y

Young, Rule of, 31, 130.

Z

Zach, 36, 44, 54-56, 134 ; Maskelyne Experiment Calculated by, 54 ; Finds Attraction of Mount Mimet, 56.
Zanotti-Bianco, 44, 54, 56, 106, 112, 113, 123, 127.

ADDENDUM

Page 32. [*Faye* (146½) *has calculated the diminution in the attraction according to his formula (see note on p. 31), and finds it to be the $\frac{1}{1187}$th part, which is not far from that resulting from the experiment. His calculation can also be stated in the following way : taking no account of the attraction of the plateau, the observed pendulum lengths reduced to sea-level by Saigey are at*

L' Isle de l' Inca	990.935 *mm.*	
Quito	991.009	"
Difference	.074	"

which difference is of the order of the errors of the observations. See this volume, p. 130, Helmert (148, *vol.* 2, *chap.* 3), *and Zanotti-Bianco* (148½, *pt.* 1, *chap.* 8, *and pt.* 2, *p.* 182).]

160

THE END